Combating Bad Weather Part II: Fog Removal from Image and Video

Synthesis Lectures on Image, Video, and Multimedia Processing

Editor
Alan C. Bovik, *University of Texas, Austin*

The Lectures on Image, Video and Multimedia Processing are intended to provide a unique and groundbreaking forum for the world's experts in the field to express their knowledge in unique and effective ways. It is our intention that the Series will contain Lectures of basic, intermediate, and advanced material depending on the topical matter and the authors' level of discourse. It is also intended that these Lectures depart from the usual dry textbook format and instead give the author the opportunity to speak more directly to the reader, and to unfold the subject matter from a more personal point of view. The success of this candid approach to technical writing will rest on our selection of exceptionally distinguished authors, who have been chosen for their noteworthy leadership in developing new ideas in image, video, and multimedia processing research, development, and education.

In terms of the subject matter for the series, there are few limitations that we will impose other than the Lectures be related to aspects of the imaging sciences that are relevant to furthering our understanding of the processes by which images, videos, and multimedia signals are formed, processed for various tasks, and perceived by human viewers. These categories are naturally quite broad, for two reasons: First, measuring, processing, and understanding perceptual signals involves broad categories of scientific inquiry, including optics, surface physics, visual psychophysics and neurophysiology, information theory, computer graphics, display and printing technology, artificial intelligence, neural networks, harmonic analysis, and so on. Secondly, the domain of application of these methods is limited only by the number of branches of science, engineering, and industry that utilize audio, visual, and other perceptual signals to convey information. We anticipate that the Lectures in this series will dramatically influence future thought on these subjects as the Twenty-First Century unfolds.

Combating Bad Weather Part II: Fog Removal from Image and Video
Sudipta Mukhopadhyay and Abhishek Kumar Tripathi
2015

Combating Bad Weather Part I: Rain Removal from Video
Sudipta Mukhopadhyay and Abhishek Kumar Tripathi
2014

Image Understanding Using Sparse Representations
Jayaraman J. Thiagarajan, Karthikeyan Natesan Ramamurthy, Pavan Turaga, and Andreas Spanias
2014

Contextual Analysis of Videos
Myo Thida, How-lung Eng, Dorothy Monekosso, and Paolo Remagnino
2013

Wavelet Image Compression
William A. Pearlman
2013

Remote Sensing Image Processing
Gustavo Camps-Valls, Devis Tuia, Luis Gómez-Chova, Sandra Jiménez, and Jesús Malo
2011

The Structure and Properties of Color Spaces and the Representation of Color Images
Eric Dubois
2009

Biomedical Image Analysis: Segmentation
Scott T. Acton and Nilanjan Ray
2009

Joint Source-Channel Video Transmission
Fan Zhai and Aggelos Katsaggelos
2007

Super Resolution of Images and Video
Aggelos K. Katsaggelos, Rafael Molina, and Javier Mateos
2007

Tensor Voting: A Perceptual Organization Approach to Computer Vision and Machine Learning
Philippos Mordohai and Gérard Medioni
2006

Light Field Sampling
Cha Zhang and Tsuhan Chen
2006

Real-Time Image and Video Processing: From Research to Reality
Nasser Kehtarnavaz and Mark Gamadia
2006

MPEG-4 Beyond Conventional Video Coding: Object Coding, Resilience, and Scalability
Mihaela van der Schaar, Deepak S Turaga, and Thomas Stockhammer
2006

v

Modern Image Quality Assessment
Zhou Wang and Alan C. Bovik
2006

Biomedical Image Analysis: Tracking
Scott T. Acton and Nilanjan Ray
2006

Recognition of Humans and Their Activities Using Video
Rama Chellappa, Amit K. Roy-Chowdhury, and S. Kevin Zhou
2005

Combating Bad Weather Part II: Fog Removal from Image and Video

Sudipta Mukhopadhyay and Abhishek Kumar Tripathi

ISBN: 978-3-031-01124-5   paperback
ISBN: 978-3-031-02252-4   ebook

DOI 10.1007/978-3-031-02252-4

A Publication in the Springer series
SYNTHESIS LECTURES ON IMAGE, VIDEO, AND MULTIMEDIA PROCESSING

Lecture #17
Series Editor: Alan C. Bovik, *University of Texas, Austin*
Series ISSN
Print 1559-8136   Electronic 1559-8144

Combating Bad Weather Part II: Fog Removal from Image and Video

Sudipta Mukhopadhyay
IIT Kharagpur

Abhishek Kumar Tripathi
Uurmi Systems

SYNTHESIS LECTURES ON IMAGE, VIDEO, AND MULTIMEDIA PROCESSING #17

ABSTRACT

Every year lives and properties are lost in road accidents. About one-fourth of these accidents are due to low vision in foggy weather. At present, there is no algorithm that is specifically designed for the removal of fog from videos. Application of a single-image fog removal algorithm over each video frame is a time-consuming and costly affair. It is demonstrated that with the intelligent use of temporal redundancy, fog removal algorithms designed for a single image can be extended to the real-time video application. Results confirm that the presented framework used for the extension of the fog removal algorithms for images to videos can reduce the complexity to a great extent with no loss of perceptual quality. This paves the way for the real-life application of the video fog removal algorithm.

In order to remove fog, an efficient fog removal algorithm using anisotropic diffusion is developed. The presented fog removal algorithm uses new dark channel assumption and anisotropic diffusion for the initialization and refinement of the airlight map, respectively. Use of anisotropic diffusion helps to estimate the better airlight map estimation. The said fog removal algorithm requires a single image captured by uncalibrated camera system. The anisotropic diffusion-based fog removal algorithm can be applied in both RGB and HSI color space. This book shows that the use of HSI color space reduces the complexity further. The said fog removal algorithm requires pre- and post-processing steps for the better restoration of the foggy image. These pre- and post-processing steps have either data-driven or constant parameters that avoid the user intervention. Presented fog removal algorithm is independent of the intensity of the fog, thus even in the case of the heavy fog presented algorithm performs well. Qualitative and quantitative results confirm that the presented fog removal algorithm outperformed previous algorithms in terms of perceptual quality, color fidelity and execution time.

The work presented in this book can find wide application in entertainment industries, transportation, tracking and consumer electronics.

KEYWORDS

bad weather, image enhancement, fog, attenuation, airlight, atmospheric visibility, anisotropic diffusion, image contrast, temporal redundancy, video enhancement, outdoor vision and weather

Dedicated to

our parents

Contents

Acknowledgments

We would like to thank our institute IIT Kharagpur for providing us a venue for this endeavor. We would like to extend our deepest gratitude to Prof. P. K. Biswas, Late Prof. S. Sengupta, Prof. A. S. Dhar, Prof. S. K. Ghosh, Prof. S. Chattopadhyay, Prof. G. Saha, Prof. S. Mahapatra and Prof. S. Banerjee for their valuable suggestions, feedback and constant encouragement.

We would also like to thank the members of the Computer Vision Laboratory. Here we had the wonderful opportunity to meet some great people. They made the work enriching and entertaining: Abhishek Midya, Jayashree, Rajat, Sumandeep, Somnath, Chiranjeevi, Manish, Jatindra, Ashis, Chanukya, Kundan, Rajlaxmi, Sandip, Vijay, Nishant and Sridevi. Special thanks to Kundan Kumar who is always ready to teach and preach LaTex. We would also thank Mr. Arumoy Mukhopadhyay for his ever-helping nature.

We feel a deep sense of gratitude to our parents and family for their encouragement and support. Finally, we would like to thank the *All mighty* for full support in this project.

Sudipta Mukhopadhyay and Abhishek Kumar Tripathi
November 2014

CHAPTER 1

Introduction

From the beginning of the cinema, visual effects have received a lot of attention. In 1896, George Méliès discovered the stop trick—one of the simplest special effects. It occurs when an object is filmed, then the camera is switched off and the object is moved out of the frame and the camera is switched on back. When the recorded clip is watched again, it gives a notion of disappearance to the viewer. Until the 1990s, special effects in post production consisted of photochemical processing steps, which were costly and time consuming. Recent digitalization of images and videos brought new possibilities in the field of post-processing. Commercial software now allows easy post-processing to the user. It is important to understand that post-processing always involves a trade-off between speed, smoothness and sharpness. Conventional special effects have been widely used in the entertainment industry. It uses a great deal of manual intervention. Nowadays, however, many special effects are introduced with the help of computer graphics and computer vision techniques with less manual intervention.

1.1 VIDEO POST-PROCESSING

The term video post-processing is used in the video/film business for quality-improvement in image processing. Digital special effects like object removal and insertion of objects are one of the key techniques in video post-processing. These tasks are mostly done after complete video sequences are recorded. Therefore, they are called "post-processing."

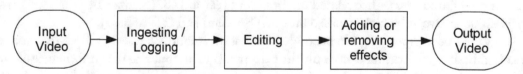

Figure 1.1: Work flow of the video post-processing.

Video post-processing begins with the raw footage and ends with a complete movie. Work-flow of the post-processing is shown in Fig. 1.1. Ingesting is a general term for capturing and importing video to a computer's hard disk. Logging is the process of identifying which shots you want to ingest. Here, scene and shot descriptions, logging notes and markers are added in the footage. Logging helps to become familiar with the footage before editing begins.

Editing involves taking the captured video along with any music or graphics and arranging these raw materials into a final edited sequence of clips. Most editors start with a rough cut, where

Figure 1.2: Visual appearance of bad weather conditions (a) rain, (b) fog and (c) snow.

they arrange all clips in the sequence. They work on fine-tuning, adjusting edit points between clips. Creating effects are more time consuming than editing.

Effects are any enhancement required to make footage more usable, such as color correction, removal or insertion of objects, animation, still or motion graphics. Once editing is finished, and effects are added or removed, video can be outputted to storage media or exported to compressor.

1.2   MOTIVATION

Car accidents are common during bad weather. It is estimated that about two million people die each year in car accidents worldwide [NHTSA, 2012]. Poor visibility is the biggest cause of accidents. On average, there are over 6,301,000 vehicle crashes each year, 24% of which (approximately 1,511,000) are weather related. Weather-related crashes are defined as those crashes that occur in adverse weather conditions (i.e., rain, snow and/or fog) or on slick pavements (i.e., wet, snowy/slushy or icy pavement). On an average, 7,130 people are killed and over 629,000 people are injured in weather-related crashes every year in U.S. (Source: 14-year averages from 1995–2008 analyzed by Noblis, based on NHTSA data) [NHTSA, 2012].

Video post-processing has been identified as an important research topic to combat bad weather. Bad weather degrades not only the perceptual quality, but also the performance of various computer vision algorithms which use visual features to accomplish tasks viz. object detection, tracking, segmentation and recognition [K. Garg and S. K. Nayar, 2007]. These computer vision algorithms assume clear day weather conditions, i.e., light reflected from an object reaches the camera unaltered. Bad weather such as rain, snow, fog and mist introduces atmospheric particles into the environment. These particles scatter the light before reaching the camera. Hence, bad weather adversely affects the performance of these computer vision algorithms. Thus, there is a requirement of post-processing that can reduce the poor visibility of bad weather and enhance the video for further processing of computer vision algorithms. Figure 1.2 shows the adverse effects of the bad weather. To design an outdoor vision system that is robust to bad weather, one needs to model their visual effects and develop algorithms to compensate.

Table 1.1: Weather conditions and associated particle types and sizes.

Condition	Particle type	Radius (μm)
Air	Molecule	10^{-4}
Haze	Aerosol	10^{-2}-1
Fog	Water droplet	1-10
Cloud	Water droplet	1-10
Rain	Water droplet	10^2-10^4

Weather conditions differ mainly in the types and sizes of the particles present in the atmosphere [S. G. Narasimhan and S. K. Nayar, 2002]. The sizes of these atmospheric particles are given in Table 1.1). When the particles are small (less than 10 μm), they float in the atmosphere and are referred to as steady (viz. fog, mist and haze) bad weather condition [S. G. Narasimhan and S. K. Nayar, 2002].

Haze is made of aerosol that is a dispersed system of particles suspended in the gas. Haze particles are larger than air molecules but smaller than fog droplets. Haze often occurs when dust and smoke particles accumulate in relatively dry air which gives the sky a cloudy appearance [S. G. Narasimhan and S. K. Nayar, 2002]. Under certain conditions of weather, when pollutants and smoke are not able to disperse, they cling together to form a hazy cloud at a low level. This obscures normal vision to a great extent. This results in the visual effect of the loss of contrast in the subject, due to the effect of light scattering through the haze particles. Fog evolves when some of the haze particles grow bigger by condensation. This transition is gradual, and the intermediate state is referred as mist. A distinction between fog and haze lies in the greatly reduced visibility induced by the former.

Everyone has heard of the horrific chain reaction accidents which occur in fog. Fog causes accidents because the driver cannot see far ahead. Foggy conditions affect the perceptual judgements of speed and distance. This effect is due to the lack of contrast [Tripathi et al., 2011]. Objects are seen based on the difference between the object brightness and background. Fog reduces the contrast significantly, causing the objects to become less distinct [Tripathi et al., 2011]. The cause of lower contrast is the same as the scattering effects of rain. Fog is produced by the suspension of very fine water droplets in the atmosphere. Smaller water droplets cause more scattering and result in loss of contrast.

CHAPTER 2

Analysis of Fog

2.1 OVERVIEW

Evaporated water molecules float in the air. As they cool down, condensation forms small water droplets. When the size of droplets are between $10^{-2} - 1$ μm, they float in the atmosphere and reduce visibility. This atmospheric condition is called as haze. When the droplet size increases further (1–10 μm), the floating droplets reduce the visibility further, and this condition is called fog. In foggy weather, the light is scatted by the water droplets. Therefore, the quantity of light traveling in a particular direction gets attenuated on the way. On the other hand, light rays traveling in all possible directions, also contributing to that particular direction of interest due to scatter [K. Garg and S. K. Nayar, 2007]. The rays reflected by the objects' surfaces are not only attenuated by the suspended water droplets but also blended with the light from the scatter, referred to as airlight, when they reach the destination. Therefore, the quality of the image captured in foggy weather is seriously degraded. The goal of the fog removal algorithm is to recover color and contrast details of the scene. Clear day images have more contrast than foggy images. Enhancement of the foggy image is a challenge due to the complexity in recovering luminance and chrominance while maintaining the color fidelity. During the enhancement of foggy images, it should be kept in mind that over enhancement leads to saturation of the pixel value. Thus, enhancement should be bounded by some constraints to avoid saturation of the image and preserve the color fidelity.

2.1.1 FRAMEWORK

It is noted that degradation of image is the function of the amount of fog present between the camera and the object. Assuming uniform fog density, it can be said that the degradation due to fog is a function of distance between the camera and the object. A generic framework for the fog removal is shown in Fig. 2.1. Removal of fog requires the estimation of image depth at

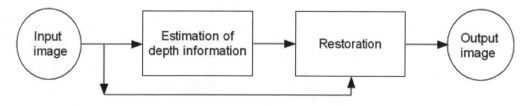

Figure 2.1: Framework for foggy image restoration.

every pixel. If we have just a single foggy image, then estimation of the depth information is an under-constrained problem. The estimation of depth using stereo imaging technique requires two images. Therefore, many methods have been proposed which use multiple images [S. G. Narasimhan and S. K. Nayar, 2002; Schechner et al., 2001; S. G. Narasimhan and S. K. Nayar, 2003]. But these methods cannot be applied on images acquired using simple uncalibrated single camera system. Hence, many attempts have been made to remove fog using a single image. To refine the estimated depth, these algorithms use some assumptions or prior knowledge. The pseudo depth can also be derived in terms of airlight map, transmission map or depth map. Some of the reported algorithms have estimated the depth information by using the scene properties. These scene properties can be shading function, human visual model or contrast-based cost function. Once depth information is estimated, it is easier to restore the image using the fog model (detailed in Section 4).

CHAPTER 3

Dataset and Performance Metrics

3.1 FOGGY IMAGES AND VIDEOS

To work on the fog removal from images and videos, we need some foggy images and videos. For extensive evaluation of fog removal algorithms at varying degrees of fog, we also need the clear images with depth information of the scene so that foggy images of varying level of degradation can be synthesized. As it is difficult to get images with depth information at every pixel in real life, most of the tests are performed with the help of no-reference metrics. Some of the examples of foggy images and videos are shown in Fig. 3.1(a)-(h). There are a number of such sites given as follows.

1. **Webpage: Dehazing Comparison** (http://johanneskopf.de/publications/deep_photo/dehazing/index.html)

2. **Webpage: Single Image Dehazing** (http://www.cs.huji.ac.il/~raananf/projects/defog/index.html)

3. **Webpage: Single Image Fog Removal Using Anisotropic Diffusion** (http://www.ecdept.iitkgp.ernet.in/web/faculty/smukho/docs/fog_removal/fog_diff.html)

4. **Webpage: Video Fog Removal** (http://www.ecdept.iitkgp.ernet.in/web/faculty/smukho/docs/fog_video/fog_video.html)

5. **Webpage: Single Image Visibility Restoration Comparison** (http://perso.lcpc.fr/tarel.jean-philippe/visibility/)

6. **Webpage: SID - Single Image Dehazing** (http://cgm.technion.ac.il/Computer-Graphics-Multimedia/Undergraduate-Projects/2009/SingleImageDehazing/ProjectWeb/)

7. **Webpage: Dehazing using Color-Lines** (http://www.cs.huji.ac.il/~raananf/projects/dehaze_cl/results/)

(a)

(b)

(c)

(d)

Figure 3.1: Video sequences useful for the simulation experiment with fog removal algorithms: (a) 'pumpkin'[R. Fattal, 2008]; (b) 'cliff'; (c) 'Yosemite 1'[J. P. Tarel and N. Hautiere, 2009], (d) 'Manhattan 2'[He et al., 2009]. *Continues.*

(e)   (f)

Figure 3.1: *Continued.* Video sequences useful for the simulation experiment with fog removal algorithms (e) 'landscape'[He et al., 2009]; (f) 'train'[He et al., 2009].

3.2 PERFORMANCE METRICS

For quantitative evaluation of the performance of any fog removal algorithm the performance metric is required. In this section, a few important metrics are presented for this purpose. Performances of fog removal algorithms for images are compared in terms of contrast gain (C_{gain}) [Economopoulosa et al., 2010] and percentage of a number of saturated pixels (σ) [Tarel et al., 2008]. For the performance of video fog removal algorithms apart from contrast gain (C_{gain}) [Economopoulosa et al., 2010] and percentage of the number of saturated pixels (σ) [Tarel et al., 2008], computation time (t_{comp}), root mean square (rms) error, and perceptual quality metric (PQM) [Wang et al., 2002] are used.

3.2.1 CONTRAST GAIN (C_{gain})

Contrast gain is described as mean contrast difference between de-foggy and foggy image. If $\bar{C}_{I_{def}}$ and $\bar{C}_{I_{fog}}$ are mean contrast of foggy image and de-foggy (restored) image, respectively, then contrast gain is defined as

$$C_{gain} = \bar{C}_{I_{def}} - \bar{C}_{I_{fog}} \tag{3.1}$$

Let an image of size $M \times N$ is denoted by $I(x, y)$. Then, mean contrast is expressed as

$$\bar{C}_I = \frac{1}{MN} \sum_{y=0}^{N-1} \sum_{x=0}^{M-1} C(x, y) \tag{3.2}$$

where

$$C(x, y) = \frac{s(x, y)}{m(x, y)} \tag{3.3}$$

where

$$m(x, y) = \frac{1}{(2p + 1)^2} \sum_{k=-p}^{p} \sum_{l=-p}^{p} I(x + k, y + l) \tag{3.4}$$

$$s(x, y) = \frac{1}{(2p + 1)^2} \sum_{k=-p}^{p} \sum_{l=-p}^{p} |I(x + k, y + l) - m(x, y)| \tag{3.5}$$

It is known that clear day images have more contrast than images plagued by fog. Hence, contrast gain for all the fog removal algorithms should be positive. Here, contrast gain metric differs from that introduced by Economopoulosa et al. [2010]. L_2 norm is replaced by L_1 norm in eqn.(3.5). This modification makes contrast gain scale invariant. Higher value of contrast gain indicates better performance of the fog removal algorithm.

3.2.2   PERCENTAGE OF THE NUMBER OF SATURATED PIXELS (σ)

High value of contrast gain means good performance. However, contrast gain should not be so high that part of the enhanced image become saturated (i.e., either completely black or white). Percentage of the number of saturated pixels (σ) is denoted as

$$\sigma = \frac{n}{M \times N} \times 100 \tag{3.6}$$

where n is the number of pixels that gets into saturation in the output image. Low value of the number of saturated pixels (σ) indicates good performance of the fog removal algorithm.

3.2.3   COMPUTATION TIME

Computation time (t_{comp}) is the time taken by the algorithm to remove fog from an image or a video frame. Obviously, t_{comp} depends upon the size of the image. Low value of t_{comp} means fast algorithm.

3.2.4   ROOT MEAN SQUARE (RMS) ERROR

For fog removal of videos, the computation time is crucial from the application point of view. Hence, modifications are suggested to speed up the algorithm at the cost of the accuracy. Degradation in the quality in restored image using such modifications with respect to the original algorithm can be measured in terms of root mean square (rms) error. Root mean square error between the reference (restored) image (I_1) and a restored image (I_2) of size $M \times N$ is given as

$$rms = \sqrt{\frac{(I_1 - I_2)^2}{M \times N}} \tag{3.7}$$

3.2.5   PERCEPTUAL QUALITY METRIC (PQM)

For the measurement of the distortion due to video coding, various performance metrics are suggested in the literature. Wang et al. [Wang et al., 2002] proposed a no-reference metric for judging the image quality reconstructed from the block DCT space to take into account visible blocking and blurring artifacts. This metric computes (i) average horizontal and vertical discontinuities at corresponding 8×8 block boundaries, (ii) activity in images expressed by deviations of average horizontal and vertical gradients from their respective average block discontinuities and (iii) the average number of zero-crossings in these gradient spaces. Subsequently, these factors are combined in the form of a nonlinear function to produce the resulting value of the metric

$$PQM = \alpha + \beta B^{\gamma_1} A^{\gamma_2} Z^{\gamma_3} \tag{3.8}$$

where $\alpha, \beta, \gamma_1, \gamma_2$ and γ_3 are model parameters that were estimated with the subjective test data as described by Wang et al [Wang et al., 2002]. B is the average blockiness, estimated as the average differences across horizontal and vertical block boundaries. A is the average absolute difference among in-block image samples and Z is the zero-crossing rate. According to C. Rouvas-Nicolis and G. Nicolis [2007] the PQM value should be close to 10 for best perceptual quality.

CHAPTER 4

Important Fog Removal Algorithms

Under fog weather conditions, contrast and color of the images are degraded. The degradation increases with the increase in distance between the camera and the object. Initial works in fog removal algorithms are based on the contrast enhancement and do not assume any fog degradation model. Thus, these algorithms can be categorized in the enhancement-based and restoration-based approaches.

4.1 ENHANCEMENT-BASED METHODS

The most commonly used methods are histogram equalization and its variants [R. C. Gonzalez and R. E. Woods, 1992]. However, these methods do not always maintain color fidelity. There are also other enhancement-based approaches like unsharp masking [R. C. Gonzalez and R. E. Woods, 1992], Retinex theory [Jobson et al., 1997] and wavelet [R. C. Gonzalez and R. E. Woods, 1992]. All enhancement-based methods have a problem in preserving color fidelity. These algorithms are designed for images whose properties are roughly constant across the image. One of the main characteristics of degradation caused by fog is that the local image contrast depends strongly on the distance from the camera. Some of these enhancement-based methods may be applied locally but at the cost of the loss of low spatial frequencies [J. P. Oakley and B. L. Satherley, 1998]. As the enhancement-based techniques fail to restore adequate quality of the foggy image, a considerable amount of research has been directed towards the knowledge of the scattering phenomena and design of the fog model. It will be discussed in details in the next section.

4.2 RESTORATION-BASED METHODS

These methods use imaging models to estimate the pattern of the image degradation, and the generated fog degradation model is used to recover the scene contrast. These methods require extra information about the imaging environment and provide a better result in comparison with the enhancement-based methods. This extra information can be in terms of estimation of scene distance, relation between shading and transmission function [R. Fattal, 2008], some heuristic assumption or pertinent scene properties.

Two fundamental phenomena which cause loss of visibility are attenuation and airlight [S. G. Narasimhan and S. K. Nayar, 2000]. Light beam coming from a scene point gets attenuated by scattering due to atmospheric particles. This phenomenon is termed as attenuation that reduces contrast in the scene. Light coming from the different sources are scattered towards camera leading to a shift in color. This phenomenon is termed as airlight. Both attenuation and airlight increase with the distance of the camera from the object.

Fog attenuation is represented as

$$I_{att}(x, y) = I_0(x, y)e^{-kd(x,y)} \tag{4.1}$$

where $I_{att}(x, y)$ is the attenuated image intensity (gray level or RGB color components) at pixel (x, y) in presence of fog and $I_0(x, y)$ is the image intensity in absence of fog (i.e., fog-free image or scene radiance), k is the extinction coefficient and $d(x, y)$ is the distance of the scene point from the viewer or the camera.

Airlight is represented as

$$A(x, y) = I_\infty(1 - e^{-kd(x,y)}) \tag{4.2}$$

where I_∞ is the global atmospheric constant. It is also called as sky intensity. According to the Koschmieder's law [Schechner et al., 2001; S. G. Narasimhan and S. K. Nayar, 2000; J. P. Tarel and N. Hautiere, 2009], the effect of fog on pixel intensity is represented as

$$I(x, y) = I_{att}(x, y) + A(x, y) \tag{4.3}$$

where $I(x, y)$ is the observed image intensity at pixel (x, y).

The Koschmieder's law is represented as

$$I(x, y) = I_0(x, y)e^{-kd(x,y)} + I_\infty(1 - e^{-kd(x,y)}) \tag{4.4}$$

where in the right-hand side, first term is the direct attenuation component and the second term is the airlight [J. P. Tarel and N. Hautiere, 2009]. When atmosphere is homogenous, the transmission map can be expressed as

$$t(x, y) = e^{-kd(x,y)} \tag{4.5}$$

Hence, a fog model can be rewritten as

$$I(x, y) = I_0(x, y)t(x, y) + I_\infty(1 - t(x, y)) \tag{4.6}$$

Brief comparison of various restoration-based prior art fog removal algorithms is shown in Table 4.1. It is shown in eqn.(4.4) that visibility of an image in the presence of fog depends on the depth. Stereoscopic depth estimation requires multiple images. Hence, initial methods use multiple images for restoration. Because of the convenience of use, later many algorithms are proposed using a single image. These algorithms estimate the depth information using some assumption or prior knowledge. Hence, restoration-based algorithms can be categorized in multiple- and single-image restoration techniques.

Table 4.1: Comparison of restoration-based methods for fog removal

Method	Number of input image(s)	Assumptions	Image type
Oakley et al [1998]	Multiple	Knowledge of scene depth	Gray
Schenchner et al [2001]	Multiple	Light scattered by atmospheric particles is partially polarized	Color & gray
Narasimhan et al [2002]	Multiple	Uniform bad weather condition	Color & gray
Narasimhan et al [2003]	Single	Interactive	Color & gray
Oakley et al [2007]	Single	Airlight is constant throughout the image	Color & gray
Kim et al [2008]	Single	Cost function-based on human visual model	Color
Kopf et al [2008]	Single	Interactive	Color & gray
Fattal [2008]	Single	Shading and transmission functions are locally uncorrelated	Color
Tan [2008]	Single	Based on spatial regularization and maximization of local contrast	Color & gray
He et al [2009]	Single	Dark channel prior	Color & gray
Tarel et al [2009]	Single	Assumes airlight as a percentage between local standard deviation and local mean of whiteness	Color & gray
Zhang et al [2010]	Single	Under the assumption that large-scale chromaticity variations are due to transmission while small-scale luminance variations are due to scene albedo	Color & gray
Fang et al [2010]	Single	Based on blackbody theory and graph-based image segmentation	Color & gray

4.2.1   MULTIPLE IMAGE-BASED RESTORATION TECHNIQUES

When the prior information of the scene structure is known, J. P. Oakley and B. L. Satherley [1998] proposed a method based on the physical model. However, this method is restricted when the prior information of the scene geometry is not available.

In another method that uses multiple images, Schechner et al. [2001] proposed a method based on polarization of light. Polarizing filter is widely used to reduce the influence of haze in photography, though it is not sufficient to remove the effect of fog altogether. Hence, this method captures two or more images taken with different degrees of polarization in the presence of fog. The difference of images with polarization filters helps to estimate airlight. This method works instantly, without any need to wait for changes in weather conditions. Airlight increases with the distance d from the object, and it can be modeled as

$$A = I_\infty(1 - e^{-kd}) \tag{4.7}$$

where k is the extinction coefficient and I_∞ is the airlight corresponding to an object at an infinite distance. The light ray from the source to a particle causing scatter and the line of sight from the camera to the same particle define a plane of incidence. Here, airlight (A) is divided into two components, which are parallel $A^\|$ and perpendicular $A^\perp$ to this plane. The airlight degree of polarization is defined as

$$P = \frac{A^\perp - A^\|}{A^\perp + A^\|} \tag{4.8}$$

The degree of polarization depends on the size and density of aerosols. The overall intensity (I) of foggy image is the sum of airlight (A) and direct attenuation (I_{att}) as

$$I = I_{att} + A \tag{4.9}$$

When polarizing filter is oriented parallel to the plane of incidence, then

$$I^\| = \frac{I_{att}}{2} + A^\| \tag{4.10}$$

When filter is oriented perpendicular to the plane of incidence, then

$$I^\perp = \frac{I_{att}}{2} + A^\perp \tag{4.11}$$

If estimate of I_∞ and P are known, then it is easy to estimate the airlight of any point as

$$A = \frac{I^\perp - I^\|}{P} \tag{4.12}$$

and unpolarized light as

$$I = I^\| + I^\perp \tag{4.13}$$

From eqn.(4.7) and eqn.(4.12), one obtains

$$e^{-kd} = 1 - \frac{A}{I_\infty} \tag{4.14}$$

This depth information helps to remove the effect of fog from the image.

S. G. Narasimhan and S. K. Nayar [2002, 2003, 2001] proposed a method that requires multiple images of the same scene captured under different weather conditions. Change in intensity of scene points under different weather conditions provides simple constraints to detect depth discontinuities in the scene and also to compute scene structure. The brightness at any pixel is given by

$$E = I_\infty \rho e^{-kd} + I_\infty (1 - e^{-kd}) \tag{4.15}$$

where I_∞ is sky intensity and ρ is the normalized radiance. Two observed pixel values E_1 and E_2 of a scene point under two weather conditions $(k_1, I_{\infty 1})$ and $(k_2, I_{\infty 2})$ can be represented as

$$E_1 = I_{\infty 1}\rho e^{-k_1 d} + I_{\infty 1}(1 - e^{-k_1 d})$$
$$E_2 = I_{\infty 2}\rho e^{-k_2 d} + I_{\infty 2}(1 - e^{-k_2 d}) \tag{4.16}$$

Eliminating ρ from eqn.(4.16), one gets

$$E_2 = \left[\frac{I_{\infty 2}}{I_{\infty 1}} e^{-(k_2 - k_1)d} \right] E_1 + \left[I_{\infty 2} \left(1 - e^{-(k_2 - k_1)d} \right) \right] \tag{4.17}$$

Equation (4.17) shows a linear relationship between E_1 and E_2. To compute sky intensity, S. G. Narasimhan and S. K. Nayar [2001] divided the images into blocks and fitted the linear relation to the (E_2, E_1) pair of scene points. Scale depth of each scene point is represented as (from eqn.(4.17))

$$(k_2 - k_1)d = -ln\left(\frac{I_{\infty 2} - E_2}{I_{\infty 1} - E_1} \right) - ln\left(\frac{I_{\infty 1}}{I_{\infty 2}} \right) \tag{4.18}$$

This depth map helps to restore the image.

4.2.2 SINGLE IMAGE-BASED RESTORATION TECHNIQUES

In the past few years, many algorithms have been proposed for the removal of the fog using a single image. These algorithms estimate the depth information using an assumption or a prior knowledge. This prior knowledge can be estimated automatically or manually. Hence, single image restoration algorithms can be categorized as interactive or automatic restoration techniques.

Interactive single image restoration

In 2008, Kopf et al. [2008] proposed a method based on the use of a 3D model [Azevedo et al., 2008] of the scene. If the depth at each pixel is known then it is easy to remove the effects of fog by fitting an analytical model as

$$I = I_0 f(d) + I_\infty(1 - f(d)) \tag{4.19}$$

where I is the intensity of foggy image, I_0 is the original image intensity, I_∞ is the sky intensity and $f(d) = exp(-kd)$ is the depth-dependent attenuation function. To restore the image, only the estimation of I_∞ and $f(d)$ is required. In this method, estimation of I_∞ and $f(d)$ requires inputs from an expert. The depth information of a pixel may be found by registration of the image with an available 3D model.

S. G. Narasimhan and S. K. Nayar [2003] proposed an interactive method which requires depth and sky intensity information from the user. Here, the user has to provide an approximate location of the vanishing point (i.e., point at largest distance) along the direction of increasing distance in the image. Next, user has to input approximate minimum and maximum distances. Other scene points' distance can be interpolated as

$$d = d_{min} + \alpha(d_{max} - d_{min}) \qquad (4.20)$$

where $\alpha \in (0, 1)$ is fractional image distance of a pixel to the vanishing point.

These interactive methods are practically not applicable for most images as no depth information is available as metadata.

Automatic single image-based restoration technique

In the recent past, many methods have been proposed which remove fog from a single image without any user intervention. J. P. Oakley and H. Bu [2007] formulated a statistical model for scene content that gives a way of detecting the presence of airlight in an arbitrary image. Here, level of airlight is estimated under the assumption that airlight is constant throughout the image. This method involves the minimization of a scalar global cost function and does not require image region segmentation. This method is applicable to both gray and color images. Once airlight level is estimated, contrast loss can be easily corrected. This method is robust and scale invariant. Accuracy of the method has been tested using Monte Carlo simulation with a synthetic image model. However, this algorithm fails when airlight is not uniform over the image.

Kim et al. [2008] improved the method proposed by J. P. Oakley and H. Bu [2007] to make it applicable even when airlight is not uniform over the image. A cost function based on human visual model is used in luminance image to estimate the airlight. The luminance image is estimated by the fusion of R, G and B color component. The airlight map is generated using linear regression, which models the relationship between regional airlight and the coordinates of the image pixels. In order to restore the image plagued by fog, airlight map is subtracted from the foggy image. This method uses region segmentation and regions are segmented uniformly to estimate the regional contribution of airlight. For an image of varying depth, airlight contribution can be varied according to the region. Estimating the airlight for each region can reflect the depth variation within the image. This algorithm gives comparatively better results but fails to cover wide range of scene depth.

R. T. Tan [2008] proposed a method based on spatial regularization from a single color or gray scale image. Tan removed the fog by maximizing the contrast of the direct transmission while assuming a smooth layer of airlight. Here fog model is assumed as given in eqn.(4.6). Based on this model, Tan assumed that for a patch with uniform transmission t, visibility (i.e., sum of the gradient) is reduced by the fog since $t < 1$:

$$\sum_{(x,y)} \|\nabla I(x, y)\| = t \sum_{(x,y)} \|\nabla I_0(x, y)\| < \sum_{(x,y)} \|\nabla I_0(x, y)\| \qquad (4.21)$$

Then Tan estimated the transmission t in a local patch by maximizing the visibility and satisfied a constraint that intensity of $I_0(x)$ is less than the intensity of I_∞. Further, Markov random field (MRF) model [P. Perez, 1998; R. Kindermann and J. L. Snell, 1980] is used to regularize the estimate. Here restored image looks saturated and produce some halos near depth discontinuities in the scene.

R. Fattal [2008] proposed a method that is based on the independent component analysis (ICA). Fattal considered the shading and transmission signals are uncorrelated and used ICA to estimate transmission, and then inferred the color of the whole image by MRF. First, the albedo of a local patch is assumed as a constant vector R. Thus, all $I_0(x, y)$ in the patch have same direction as R. Then direction of R is estimated by ICA, by assuming that the surface shading $||I_0(x, y)||$ and the transmission $t(x, y)$ are independent in the patch. Finally, MRF model guided by input color image is applied to extrapolate the solution to the whole image. This method estimates the optical transmission in foggy scenes. Based on this estimation, the scattered light is eliminated to increase scene visibility and remove the effect fog from scene contrasts. Here restoration is based on the color information; hence, this method can not be applied for the gray image. This method fails in dense fog because dense fog is often colorless.

He et al. [2009] proposed a method based on the dark channel prior and soft matting. Dark channel prior is based on a key observation that most local patches in foggy outdoor images contain pixels that have low intensities in at least one color component. Thus, for an image, dark channel image is defined as

$$I^{dark}(x, y) = \min_{c \in \{r,g,b\}} \left(\min_{(x,y) \in \Omega} (I^c(\Omega)) \right) \tag{4.22}$$

where Ω is a local patch in the image. Then the complement of the dark channel is assumed as the coarse transmission map as

$$\tilde{t}(x, y) = 1 - \omega I^{dark}(x, y)$$
$$= 1 - \omega \min_{c \in \{r,g,b\}} \left(\min_{(x,y) \in \Omega} (I^c(\Omega)) \right) \tag{4.23}$$

where ω ($0 < \omega \leq 1$) is a weighting factor, and its value is application dependent. Then Levin's soft matting [Levin et al., 2006] algorithm is used to refine transmission map. The optimal transmission map t can be obtained by solving the following sparse linear system

$$(L + \lambda U)t = \lambda \tilde{t} \tag{4.24}$$

where U is the identity matrix, L is the matting Laplacian matrix proposed by Levin et al. [2006] and λ is a regularization parameter. Final scene radiance $I_0(x)$ is recovered by

$$I_0(x, y) = \frac{I(x, y) - I_\infty}{max(t(x, y), 0.1)} + I_\infty \tag{4.25}$$

where I_∞ is estimated with the most foggy pixel. The disadvantage of this algorithm is that, when the scene objects are nearly as bright as the atmospheric light, underlying assumption of this algorithm does not remain valid.

J. P. Tarel and N. Hautiere [2009] proposed a fast visibility restoration algorithm. This method assumes the airlight as a percentage of difference of the local mean of the image and the local standard deviation of the intensity profile. This method is based on linear operations but requires many parameters for the adjustment. The first step of image restoration is inferring atmospheric veil $A(x, y)$ (or airlight). Since visibility restoration is an ill-posed problem, Tarel et al. solved this by maximizing the contrast of the resulting image assuming that the depth map must be smooth except along the edges. Tarel et al formalized the following optimization problem with constraint $0 \leq A(x, y) \leq W(x, y)$:

$$\underset{A}{argmax} \int_{(x,y)} A(x, y) - \lambda\phi \left(||\nabla A(x, y)||^2\right) \, dx \, dy \tag{4.26}$$

where $W(x, y) = \underset{c \epsilon (r,g,b)}{min} (I(x, y))$ defined as the image of the minimal component of $I(x, y)$ for each pixel. Parameter λ controls the smoothness of the solution and ϕ is an increasing concave function. Since this optimization is computationally intensive, Tarel et al. bypassed this optimization problem with a linear operation as

$$
\begin{aligned}
A(x, y) &= max \left(min \left(pC(x, y), W(x, y)\right), 0\right) \\
\text{with } C(x, y) &= B(x, y) - median_{w \times w}(|W - B|)(x, y) \\
\text{and } B(x, y) &= median_{w \times w}(W)(x, y)
\end{aligned}
\tag{4.27}
$$

where factor p $(0 < p < 1)$ control the strength of the visibility restoration, and $w \times w$ is the spatial window size. Then restored image is obtained as

$$I_0(x, y) = \frac{I(x, y) - A(x, y)}{1 - \frac{A(x,y)}{I_\infty}} \tag{4.28}$$

The main advantage of this algorithm is its speed. The complexity of the algorithm is of the order of number of image pixels. This speed paves the way for real-time implementation of fog removal algorithm. But the restored image quality is not so good when there are discontinuities in the scene depth.

Zhang et al. [2010] proposed a method under the assumption that large-scale chromaticity variations are due to transmission map while small scale luminance variations are due to scene albedo. A nonlinear edge-preserving filter is introduced here to refine subtle transmission map incrementally while still keeping sharp transmission map distinct. This method is based on iterative bilateral filter. This algorithm gives good results. Due to the use of iterative bilateral filter this technique is computationally intensive and requires the choice of the number of parameters (viz. spatial and intensity kernels of bilateral filter and numbers of color groups) for optimal results. For a good result, these parameters need to be adjusted for each image. For images with large-scale atmosphere color-like object, this method can not calculate the transmission map correctly.

Fang et al. [2010] proposed a method based on the graph-based segmentation. Here, graph-based image segmentation is applied to segment the foggy image. Then initial transmission map is obtained according to the blackbody theory. After that, the bilateral filter is used to refine the transmission map. It is noted that for the foggy image, choice of segmentation control parameters is difficult.

CHAPTER 5

Single-Image Fog Removal Using an Anisotropic Diffusion

5.1 INTRODUCTION

In this chapter, a novel fog removal algorithm is presented. The said algorithm uses modified dark channel assumption for the estimation of initialization of airlight map and anisotropic diffusion for its refinement. The fog removal algorithm suggests pre- and post-processing steps to increase the quality of restored image. The said algorithm can be applied for color as well as grayscale images. For color images, the presented algorithm works in RGB or HSI color space. This algorithm has few number of parameters and constants. The parameters are data driven, and the value of the constants remain same irrespective of the image under consideration.

This chapter is structured as follows. In Section 5.2, the novel fog removal algorithm is explained in detail. In Section 5.3, simulation and results are discussed. Here performance of the presented algorithm is compared with the previous algorithms. Section 5.4 concludes this chapter.

5.2 FOG REMOVAL ALGORITHM[1]

According to Koschmieder law [J. P. Tarel and N. Hautiere, 2009], model of fog effect can be represented as

$$I(x, y) = I_0(x, y)e^{-kd(x,y)} + I_\infty(1 - e^{-kd(x,y)}) \tag{5.1}$$

where $I_0(x, y)$ is the image intensity at pixel location (x, y) in absence of fog, k is extinction coefficient, $d(x, y)$ is the distance of the scene point from the camera or viewer, I_∞ is global atmospheric constant or sky intensity and $I(x, y)$ is the observed image intensity at pixel location (x, y) in presence of fog.

In eqn. (5.1), in the right-hand side, the first term is the direct attenuation term and the second term is the airlight term. Attenuation is an exponential decreasing function. The contrast of the object is reduced due to attenuation and thus its visibility in scene. Airlight adds whiteness in the scene. Airlight is an increasing function of the scene point distance $d(x, y)$. If airlight is represented as $A(x, y)$ then eqn. (5.1) can be written as

---

[1]Patent application No. 1029/KOL/2011, titled as A METHOD AND SYSTEMS FOR REMOVAL OF FOG FROM THE IMAGES AND VIDEOS, applied on 03-AUG-2011. PCT Application No. PCT/IN2012/000077 dated 02-Feb2012.

$$I(x, y) = I_0(x, y) \left(1 - \frac{A(x, y)}{I_\infty}\right) + A(x, y) \tag{5.2}$$

For simulation foggy image intensity $I(x, y)$ is normalized. Fog being pure white, and the intensity of the sky is highest in the daytime, the sky intensity I_∞ can be set to 1. Hence, for the restoration of the image $I_0(x, y)$, only the information of the airlight $A(x, y)$ is required.

From eqn. (5.2), we get

$$I_0(x, y) = \frac{1}{(1 - A(x, y))} (I(x, y) - A(x, y)) \quad \text{assuming } I_\infty = 1 \tag{5.3}$$

This model is directly extended for color images by applying the same model on each of the RGB components. Here, airlight map $A(x, y)$ remains same for each of the color components.

The color image can also be represented in HSI (Hue, Saturation and Intensity) color space [R. C. Gonzalez and R. E. Woods, 1992]. It is known that the HSI model is closer to human perception. Hue is the color attribute that describes a pure color, and saturation gives a measure of the degree to which a pure color is diluted by white light. This color space decouples the intensity component from the color carrying information (hue and saturation) in a color image. It is observed that fog has no effect on the hue of the scene. The other two components' saturation and intensity are affected by fog. Hence, to restore a foggy image, processing is needed only in the saturation and intensity planes. Thus, unlike RGB color space, computation is reduced by $\frac{1}{3}$ in HSI color space. In HSI color space, the effect of fog on the intensity plane is governed by eqn. (5.2). For the saturation component, fog effect is derived as follows.

Taking minimum across color components of both side in eqn. (5.2)(assuming $I_\infty = 1$), we get

$$\min_{c\epsilon(r,g,b)} (I^c(x, y)) = \min_{c\epsilon(r,g,b)} (I_0^c(x, y)) (1 - A(x, y)) + A(x, y) \tag{5.4}$$

For the intensity component (average of R,G and B) from eqn. (5.3), we get

$$I_{0\,int}(x, y) = \frac{1}{(1 - A(x, y))} (I_{int}(x, y) - A(x, y)) \tag{5.5}$$

From eqn. (5.3) and eqn. (5.5)

$$\frac{\min\limits_{c\epsilon(r,g,b)} (I_0^c(x, y))}{I_{0\,int}(x, y)} = \frac{\min\limits_{c\epsilon(r,g,b)} (I^c(x, y)) - A(x, y)}{I_{int}(x, y) - A(x, y)}$$

or

$$(1 - S_{I_0}(x, y)) = \frac{\min\limits_{c\epsilon(r,g,b)} (I^c(x, y)) \left[1 - \frac{A(x,y)}{\min\limits_{c\epsilon(r,g,b)} (I^c(x,y))}\right]}{I_{int}(x, y) \left[1 - \frac{A(x,y)}{I_{int}(x,y)}\right]}$$

or

$$S_{I_0}(x, y) = 1 - \frac{(1 - S_I(x, y)) \left[1 - \frac{A(x,y)}{\min\limits_{c \in (r,g,b)} (I^c(x,y))} \right]}{\left[1 - \frac{A(x,y)}{I_{int}(x,y)} \right]} \tag{5.6}$$

where $S_{I_0}(x, y)$ is saturation in absence of fog and $S_I(x, y)$ is saturation of the foggy image.

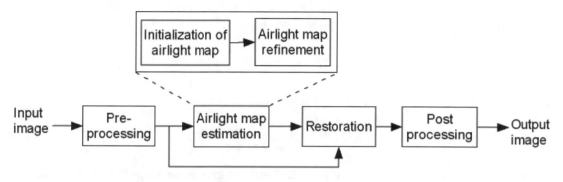

Figure 5.1: Block diagram of the fog removal algorithm.

Block diagram of the fog removal algorithm is shown in Fig. 5.1. In order to remove fog, first pre-processing (i.e., histogram equalization) is performed on the foggy image. This pre-processing increases the contrast of the image prior to the fog removal and results better estimation of airlight map. Adaptive histogram equalization is used for the pre-processing operation. It works on $L.a.b.$ color space and equalization is applied on L channel. Initial value of airlight map is estimated by black channel prior. Airlight map is refined using anisotropic diffusion. Once airlight map is obtained, image is restored using eqn. (5.3) and eqn. (5.6). Histogram stretching of the output image is performed as post-processing. Transfer function of the histogram stretching is calculated using the intensity plane. This transfer function is applied independently on each R, G, B color channels. The fog removal algorithm adopted the data-driven transfer function for the histogram stretching to avoid user intervention (discussed later). This histogram stretched image is the final restored image.

5.2.1   INITIALIZATION OF AIRLIGHT MAP

It is known that airlight map A is a scalar image which is always positive. Using minimal operator in eqn. (5.3) over RGB color space, we get

$$A(x, y) = \min_{c \in (r,g,b)} (I^c(x, y)) - \min_{c \in (r,g,b)} \left[I_0^c(x, y) (1 - A(x, y)) \right] \tag{5.7}$$

According to the dark channel prior [He et al., 2009], dark channel is denoted as the minimum intensity across red, blue and green channels over a small patch. Present dark channel assumption differs from the assumption forwarded by He et al. [2009]. This dark channel uses only color space, i.e., minimum across R, G and B channels at a particular pixel location instead of the spatial window on color space as presented earlier. This modification reduces the calculation significantly without loss of quality. Natural outdoor images are, usually, full of shadows and colorful objects (viz. green grass, trees, red or yellow plants and blue water bodies). Thus, dark channel assumption, i.e., one channel is dark in the absence of fog, is valid for these images. In the fog-free image except for sky region intensity of dark channel is low and tends to zero [He et al., 2009].

Hence, $\min\limits_{c\epsilon(r,g,b)} \left(I_0^c(x,y)\right) \approx 0$. Thus, from eqn. (5.7), we get

$$\min\limits_{c\epsilon(r,g,b)} \left(I^c(x,y)\right) \geq A(x,y) > 0 \qquad (5.8)$$

Initial estimation of airlight map $(A_0(x,y))$ is assumed as

$$A_0(x,y) = \beta \min\limits_{c\epsilon(r,g,b)} \left(I^c(x,y)\right) \qquad (5.9)$$

where β is a constant and $0 < \beta < 1$. If input image is the gray-scale image then initial estimation can be assumed as

$$A_0(x,y) = \beta I(x,y) \qquad (5.10)$$

5.2.2   AIRLIGHT MAP REFINEMENT

Airlight map is the function of the distance of the object from the camera. Different objects may be at different distance from the camera, and thus, airlight map should differ from object to object. Considering the continuity of the object, the variation in airlight map should be small except along the object edges. Hence, airlight map should have intra-region smoothing preferentially over inter-region smoothing. The anisotropic diffusion [P. Perona and J. Malik, 1990] can fulfill this requirement.

Anisotropic diffusion can be represented as

$$\frac{\partial A}{\partial t} = div\left(\alpha(x,y,t)\nabla A\right)$$
$$= \alpha(x,y,t)\triangle A + \nabla\alpha \cdot \nabla A \qquad (5.11)$$

where div is divergence operator and α is conduction coefficient. ∇ and $\triangle$ are gradient and Laplacian operators respectively. If $\alpha(x,y,t)$ is constant over time and space then eqn. (5.11) reduces to isotropic heat diffusion equation

$$\frac{\partial A}{\partial t} = \alpha\triangle A \qquad (5.12)$$

Heat equation describes the distribution of heat in the region over time. Heat equation states that if a hot body is kept inside the box of cold water, how the temperature of the body will decrease and finally attain the temperature of the surrounding water. To encourage the smoothing within a region α should be one in the interior region and zero at the edges to avoid smoothing. If $E(x, y, t)$ is the estimation of the boundaries, then according to Perona-Malik equation [P. Perona and J. Malik, 1990] the conduction coefficient should be chosen as

$$\alpha = g(\|E\|) \tag{5.13}$$

where $g(\cdot)$ is a nonnegative monotonically decreasing function with $g(0) = 1$. Thus, diffusion takes place in the interior of a region without affecting the region boundaries.

Here, $g(\cdot)$ is assumed as [P. Perona and J. Malik, 1990]

$$g(\|E\|) = e^{-\left(\frac{\|E\|}{\kappa}\right)^2} \tag{5.14}$$

where κ is a positive constant which is fixed. Hence, according to eqn. (5.11) airlight map can be estimated iteratively as follows [P. Perona and J. Malik, 1990]:

$$A^{t+1} = A^t + \lambda[\alpha \nabla A^t] \tag{5.15}$$

where λ ($0 < \lambda < 1$) is a smoothing parameter. According to Perona-Malik discrete version of eqn. (5.15) can be written as

$$\begin{aligned} A(x, y, t + 1) = A(x, y, t) + \lambda[&\alpha_N(x, y, t)\nabla_N A(x, y, t) \\ &+ \alpha_S(x, y, t)\nabla_S A(x, y, t) \\ &+ \alpha_E(x, y, t)\nabla_E A(x, y, t) \\ &+ \alpha_W(x, y, t)\nabla_W A(x, y, t)] \end{aligned} \tag{5.16}$$

where N, S, E, W are the mnemonic subscripts for North, South, East and West. Symbol ∇ indicates nearest-neighbor differences.

5.2.3   BEHAVIOR OF ANISOTROPIC DIFFUSION

Anisotropic diffusion performs smoothing of intra-region and edges remain stable over a very long time, as shown in Fig. 5.2. The terrain presented in Fig. 5.2(a) is difficult to recognize after adding noise (see Fig. 5.2(b)). After anisotropic diffusion (10 iterations) terrain is restored close to the original (see Fig. 5.2(c)). Anisotropic diffusion has preserved edges while smoothing. It is because instead of considering diffusion and edge detection as two independent processes, here both the processes interact in one single process. Here, an inhomogeneous smoothing process is applied that reduces the diffusivity at those locations that have a larger likelihood to be edges.

5.2.4   RESTORATION

Once airlight map A is estimated then each color component of foggy free image $I_0(x, y)$ can be restored as (using eqn. (5.3))

$$I_0^c(x, y) = \frac{(I^c(x, y) - A(x, y))}{(1 - A(x, y))} \tag{5.17}$$

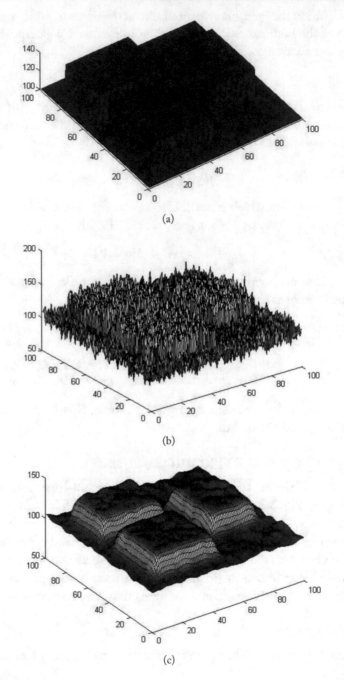

Figure 5.2: (a) Surface plot of terrain, (b) terrain embedded in noise and (c) after anisotropic diffusion at the end of 10 iterations.

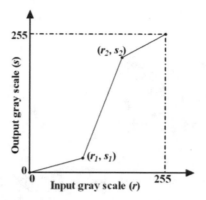

Figure 5.3: Transformation function of the histogram stretching as post-processing of the restored output of the image degraded by fog.

where $c \in (r, g, b)$. It is noted that eqn. (5.17) can be applied for gray-scale image also. For gray-scale images, the initial estimation of airlight map is discussed in eqn. (5.10). Restoration of foggy image in HSI color space can be achieved by using eqn. (5.17) for intensity component and eqn. (5.6) for saturation component, whereas the hue component remains unaltered.

5.2.5 POST-PROCESSING

It is observed that restored image have low contrast. Thus, there is a requirement of contrast enhancement as a post-processing operation. There are many choices for the contrast enhancement like histogram equalization, histogram specification and histogram stretching. It is found that histogram equalization produces a saturated output image. Due to large variations in image content, a standard reference image cannot be fixed which is required for histogram specification. Thus, to increase contrast, the algorithm uses histogram stretching. Transfer function for histogram stretching is shown in Fig. 5.3, where axis r and s represent input and output pixel values respectively, and parameters r_1, s_1, r_2 and s_2 determine the shape of the transfer function. The choice of these parameters depends on the image histogram (detailed in Section 5.3).

5.3   SIMULATION AND RESULTS

The simulation is carried out in MATLAB 7.0.4 environment. Qualitative and quantitative performance of the algorithm is compared with existing algorithms. Value of $\beta = 0.9$ for initialization and $\kappa = 30$ & $\lambda = \frac{1}{7}$ for refinement of airlight map are found to give good results for all images used in the simulation. Although anisotropic diffusion is an iterative process, it converges within ten iterations for all the images used in this experiment. In the post-processing step, the cumulative histogram of input foggy image is used to set values of r_1 and r_2. The values of r_1 and r_2 are chosen as the intensity values of 10% and 90% of the cumulative histogram, respectively.

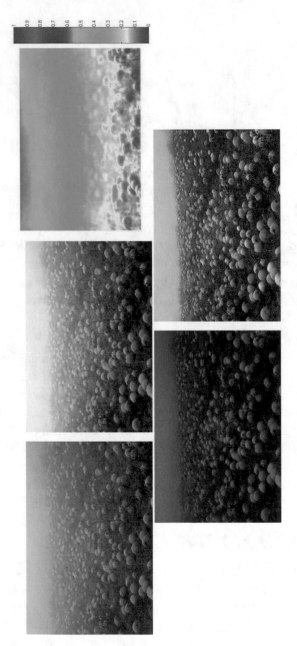

Figure 5.4: (a) Original 'pumpkins' image, (b) after pre-processing of (a), (c) corresponding airlight map estimation, (d) restored output (using presented (RGB)), (e) final restored image derived by post-processing of (d).

Figure 5.5: The first column shows the original foggy (a) 'dooars01', (d) 'dooars02', (g) 'park' and (j) 'valley' images. The second and third columns show the corresponding images restored by presented anisotropic diffusion based restoration algorithm working in RGB color space and HSI color space, respectively. *Continues.*

(j)   (k)   (l)

Figure 5.5: *Continued.* The second and third columns show the corresponding images restored by presented anisotropic diffusion-based restoration algorithm working in RGB color space and HSI color space, respectively.

The values of s_1 and s_2 are chosen as 5% and 95% of the output intensity range, i.e., [0 255]. Here, values of r_1, r_2, s_1, and s_2 determine the shape of the post processing transfer function. The enhancement parameters being data driven, avoids the need of the user intervention. Use of histogram equalization as pre-processing step and histogram stretching as a post-processing step is found to improve the quality of restoration.

Result of airlight map estimation of the image 'pumpkins' is shown in Fig. 5.3. It is observed that in foggy image estimated airlight map depends upon the distance of the scene points from the camera. Estimated airlight map is discontinuous across edges and smooth within objects. The airlight map is a gray image. For better visualization, it is shown in pseudo-color. Result of intermediate steps of the anisotropic diffusion-based algorithm working in RGB color space are shown in Fig. 5.4. The intermediate result helps to appreciate the contribution of each step. Pre-processing enhances the contrast that helps to estimate more accurate airlight map. A better estimation of airlight map results in better restoration. The post-processing provides a visually pleasant output image.

Results of the presented algorithm using RGB and HSI color space are shown in Fig. 5.5. Results show that the presented algorithm (using RGB and HSI color space) restores foggy images well with acceptable visual quality. Qualitative performance of presented algorithm is compared with the other prior art algorithms. Results are shown in Figs. 5.6, 5.7, 5.8, 5.9 and 5.10. Results show that the presented algorithm restores foggy images better than the existing algorithms. For quantitative analysis, C_{gain} and σ are measured for full resolution images. It is known

Figure 5.6: (a) Original foggy image 'ny17'. Output image after restoration by the algorithm (b) Fattal, (c) He et al., (d) Tarel et al., (e) Zhang et al., the anisotropic diffusion algorithm in (f) RGB color space and (g) HSI color space.

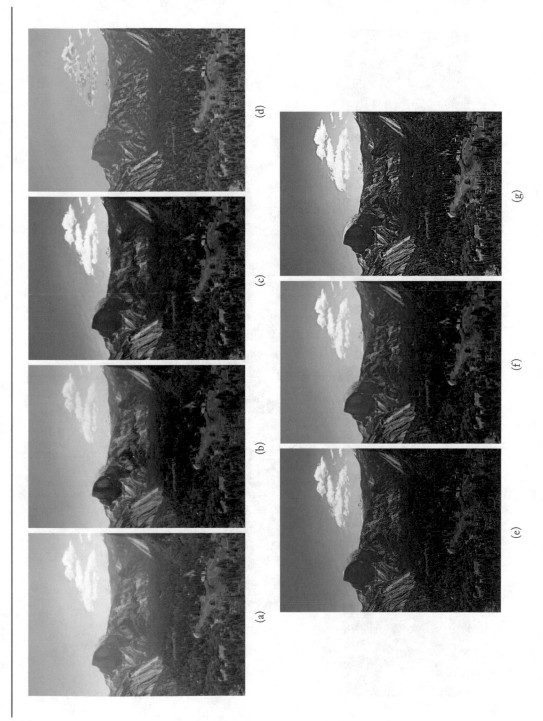

Figure 5.7: (a) Original foggy image 'y01'. Output image after restoration by the algorithm (b) Fattal, (c) He et al., (d) Tarel et al., (e) Zhang et al., the anisotropic diffusion algorithm in (f) RGB color space and (g) HSI color space.

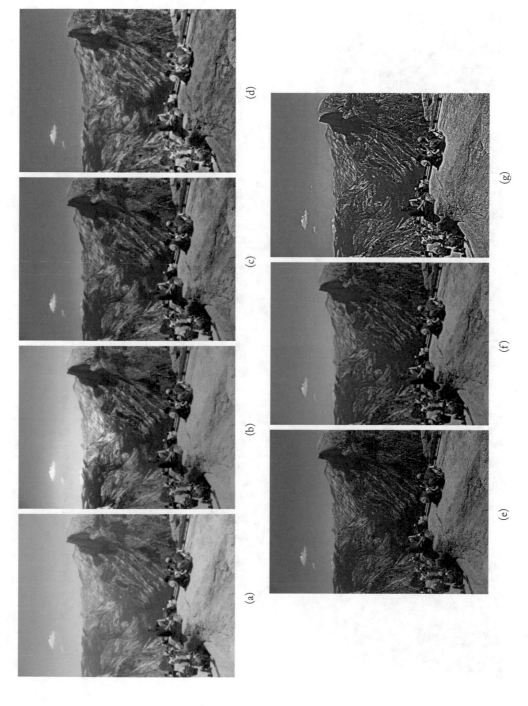

Figure 5.8: (a) Original foggy image 'y16'. Output image after restoration by the algorithm (b) Fattal, (c) He et al., (d) Tarel et al., (e) Zhang et al., the anisotropic diffusion algorithm in (f) RGB color space and (g) HSI color space.

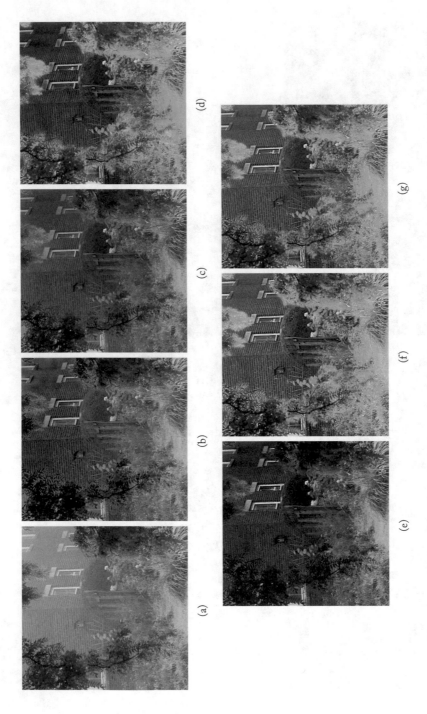

Figure 5.9: (a) Original foggy image 'house'. Output image after restoration by the algorithm (b) Fattal, (c) He et al., (d) Tarel et al., (e) Zhang et al., the anisotropic diffusion algorithm in (f) RGB color space and (g) HSI color space.

Figure 5.10: (a) Original foggy image 'stadium'. Output image after restoration by the algorithm (b) Fattal, (c) He et al., (d) Tarel et al., (e) Zhang et al., the anisotropic diffusion algorithm in (f) RGB color space and (g) HSI color space.

Table 5.1: Contrast gain (C_{gain}) and percentage of saturated pixels (σ) produced by presented and competing fog removal algorithms for different images

Image	Method													
	Fattal		He et al.		Tarel et al.		Zhang et al.		Presented (RGB)		Presented (HSI)	
	C_{gain}	σ	C_{gain}	σ	C_{gain}	σ	C_{gain}	σ	C_{gain}	σ	C_{gain}	σ
'ny17'	0.102	2.0058	0.093	0.2215	0.116	0.0011	0.156	0.0007	0.165	0.0001	**0.178**	**0.0000**
'y01'	0.061	0.1155	0.086	1.0084	0.084	0.0011	0.108	0.0001	**0.111**	**0.0000**	0.111	0.0000
'y16'	0.039	0.3203	0.068	0.1738	0.091	0.0004	0.117	0.0007	0.146	**0.0003**	**0.254**	0.0683
'house'	0.157	4.4918	0.036	**0.0000**	0.082	**0.0000**	0.157	0.2469	0.162	**0.0000**	**0.169**	**0.0000**
'stadium'	0.117	0.5312	0.139	0.3349	0.086	**0.0000**	0.183	0.0232	0.172	**0.0000**	**0.219**	**0.0000**

Table 5.2: Contrast gain (C_{gain}) and percentage of saturated pixels (σ) produced by presented (excluding pre and post processing steps) and competing fog removal algorithms for different images

Image	Method												
	Fattal		He et al.		Tarel et al.		Zhang et al.		Presented (RGB)		Presented (HSI)	
	C_{gain}	σ	C_{gain}	σ	C_{gain}	σ	C_{gain}	σ	C_{gain}	σ	C_{gain}	σ
'ny17'	0.102	2.0058	0.093	0.2215	0.116	0.0011	0.156	0.0007	0.1558	0.0000	0.1606	0.0000
'y01'	0.061	0.1155	0.086	1.0084	0.084	0.0011	0.108	0.0001	0.1097	0.0000	0.1083	0.0000
'y16'	0.039	0.3203	0.068	0.1738	0.091	0.0004	0.117	0.0007	0.1189	0.0000	0.1691	0.0000
'house'	0.157	4.4918	0.036	0.0000	0.082	0.0000	0.157	0.2469	0.0621	0.0000	0.1555	0.0000
'stadium'	0.117	0.5312	0.139	0.3349	0.086	0.0000	0.183	0.0232	0.1214	0.0000	0.1666	0.0000

that an over-saturated image may give high contrast gain (C_{gain}) but produces a large number of saturated pixels (σ). Hence, two performance metrics C_{gain} and σ are analyzed together. High value of C_{gain} and low value of σ indicate better performance of the algorithm. For the simulation of contrast gain 5×5 window ($p = 2$) is chosen. Results for C_{gain} and σ are shown in Table 5.1. Results show that contrast gain (C_{gain}) for presented algorithm (using RGB and HSI color space) is higher than the existing algorithms. Percentage of saturated pixels (σ) value for presented algorithm is lower than R. Fattal [2008], He et al. [2009], Zhang et al. [2010] and J. P. Tarel and N. Hautiere [2009]. According to the contrast gain these algorithms can be arranged in descending order as: Presented (HSI), Presented (RGB), Zhang et al., Tarel et al., He et al. and Fattal. According to the σ these algorithms can be arranged in ascending order as: Presented (RGB), Presented (HSI), Zhang et al., Tarel et al., He et al. and Fattal.

For each of the previous algorithms, restored image looks dim. For better visualization, all algorithms use pre- and post-processing. If pre- and post-processing steps are removed from the presented fog removal algorithm (in RGB and HSI color space) then C_{gain} and σ will decrease but maintain the same trend of the result. The results are shown in Table 5.2. It can be observed that presented algorithm (RGB and HSI color space) still provides favorable results in comparison with other techniques.

It is noted that extinction coefficient (k) controls the amount of fog in images. Two sets of foggy images are generated for two images (white flower and forest) with known depth map by varying extinction coefficient in eqn. (5.1). A higher value of k produces a higher amount of fog in the image. These images, plagued by different concentration of fog, are restored by present and existing algorithms. Quantitative comparison of the competing techniques is shown in Table 5.3 and Table 5.4. Results confirm that the efficacy of the said restoration algorithm (RGB & HSI) is independent of the fog density. For varying fog density, the presented technique (RGB & HSI) produces high value of C_{gain} and low value of σ. Presented algorithm outperformed other existing algorithms with a significant margin. Qualitative results are shown in Fig. 5.11 and Fig. 5.12. Presented algorithm (RGB and HSI color space) restores images better than other algorithms with acceptable image quality. More results can be found at `http://www.ecdept.iitkgp.erne t.in/web/faculty/smukho/docs/fog_removal/fog_diff.html`.

Novelties of the said algorithm are in modified assumption for estimation of dark channel, use of anisotropic diffusion for refinement of airlight map and pre- and post-processing operations for better restoration. In the presented algorithm for dark channel estimation, the minimum operator is applied only on the color components of the pixel at a particular instant instead of the previous assumption where the operator is applied on all the color components of the pixels within a chosen window. This new assumption reduces the execution time. Anisotropic diffusion further reduces the computation in comparison with bilateral filter used in the prior work [Zhang et al., 2010] (detailed in Section 6.4). In anisotropic diffusion, the kernel used for estimation of edges is Laplacian operator, where coefficients of the kernel are fixed. However, in case of the bilateral filter the kernel coefficients are image dependent. Zhang et al. [2010] used iterative bilateral

Table 5.3: Contrast gain (C_{gain}) and percentage of saturated pixels (σ) of the image 'white flower' plagued by various intensities of fog, restored by the competing algorithms

k	Fattal		He et al.		Tarel et al.		Zhang et al.		Presented(RGB)		Presented(HSI)	
	C_{gain}	σ	C_{gain}	σ	C_{gain}	σ	C_{gain}	σ	C_{gain}	σ	C_{gain}	σ
0.2	0.167	0.1132	0.202	0.0326	0.078	**0.0000**	0.153	**0.0000**	0.202	**0.0000**	**0.247**	0.0025
0.5	0.174	0.1102	0.211	0.0426	0.057	**0.0000**	0.125	**0.0000**	0.237	**0.0000**	**0.285**	0.0026
0.8	0.171	0.1281	0.248	0.0852	0.045	**0.0000**	0.109	**0.0000**	0.272	**0.0000**	**0.287**	0.0034
1.2	0.177	0.2010	0.265	0.1785	0.034	**0.0000**	0.086	**0.0000**	0.279	**0.0000**	**0.282**	0.0074
1.5	0.164	0.2263	0.272	0.2084	0.031	**0.0000**	0.073	**0.0000**	0.281	**0.0000**	**0.283**	0.0127

Table 5.4: Contrast gain (C_{gain}) and percentage of saturated pixels (σ) of the image 'forest' plagued by various intensities of fog, restored by the competing algorithms

k	Method												
	Fattal		He et al.		Tarel et al.		Zhang et al.		Presented(RGB)		Presented(HSI)	
	C_{gain}	σ	C_{gain}	σ	C_{gain}	σ	C_{gain}	σ	C_{gain}	σ	C_{gain}	σ
0.2	0.088	0.0120	0.109	0.0006	0.072	**0.0000**	0.128	0.0002	0.128	**0.0000**	**0.138**	0.0002
0.5	0.082	0.0170	0.117	0.0003	0.062	**0.0000**	0.143	0.0002	0.163	**0.0000**	**0.171**	0.0003
0.8	0.095	0.0171	0.129	0.0003	0.052	**0.0000**	0.114	0.0001	0.166	**0.0000**	**0.171**	0.0004
1.2	0.094	0.0201	0.132	0.0041	0.040	0.0001	0.088	**0.0000**	0.161	**0.0000**	**0.172**	0.0004
1.5	0.088	0.0192	0.135	0.0007	0.031	0.0001	0.066	0.0001	0.158	**0.0000**	**0.169**	0.0004

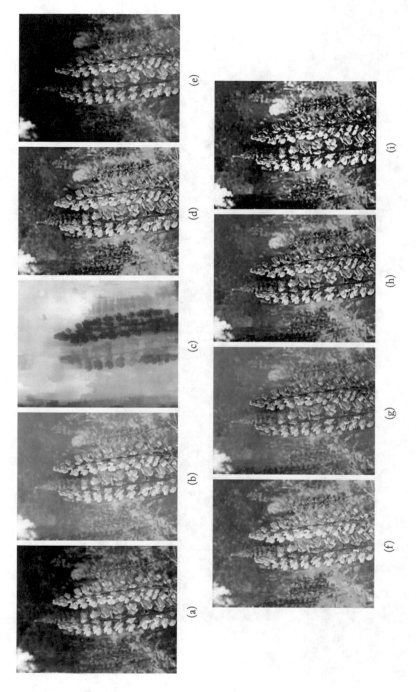

Figure 5.11: (a) Original 'white flower' image, (b) foggy 'white flower' image, and (c) corresponding depth image of 'white flower' image. Foggy 'white flower' Image resorted by (d) Fattal, (e) He et al., (f) Tarel et al., (g) Zhang et al., (h) presented anisotropic diffusion-based algorithm applied in RGB color space and (i) presented anisotropic diffusion based algorithm applied in HSI color space.

Figure 5.12: (a) Original 'forest' image, (b) foggy 'forest' image ($k = 1.5$), and (c) corresponding depth image of 'forest' image. Foggy 'forest' Image resorted by (d) Fattal, (e) He et al., (f) Tarel et al., (g) Zhang et al., (h) presented anisotropic diffusion based algorithm applied in RGB color space and (i) presented anisotropic diffusion based algorithm applied in HSI color space.

filter, where in each iteration kernel coefficients are computed, which increases the complexity. Presented algorithm requires only three parameters as input and rest of the parameters are data driven. The value of these three parameters (β, λ, κ) remains constant for all the images used in this chapter. This decision makes the presented algorithm automatic and easy to implement and use in real life.

5.4 CONCLUSION

In this chapter, a novel, efficient fog removal algorithm is presented. This algorithm uses anisotropic diffusion for estimation of airlight. The fog removal algorithm does not require user intervention and can be applied for color as well as gray image. For the color image, the said restoration algorithm can be applied using both RGB and HSI color space. Results show that the anisotropic diffusion-based algorithm performs well in comparison with other existing algorithms. Even in the case of heavy fog, the presented algorithm performs well and the quality of the restored image is independent of the density of fog present in the image. In order to evaluate the performance, contrast gain and percentage of the number of saturated pixels are used. Results confirm that presented algorithm beats the competing techniques with a significant margin. This algorithm can be used as a pre-processing for various algorithms such as object detection, tracking and segmentation.

CHAPTER 6

Video Fog Removal Framework Using an Uncalibrated Single Camera System

6.1 INTRODUCTION

In foggy weather, driving a vehicle is more difficult than in normal weather conditions. Fog reduces the visibility and reduction of visibility increases the risk of an accident. A real-time fog removal algorithm is now needed. At present, there is no algorithm that is specifically designed for video fog removal. However, in prior work [J. P. Tarel and N. Hautiere, 2009] it is mentioned that the fog removal algorithm can be extended to video by applying the image restoration algorithm on each frame. It is found that frame wise restoration is not feasible for real-time application due to the computational burden of the reported algorithms.

This chapter aims to combat the fog effect from videos by utilizing the temporal correlation among neighboring video frames. If objects in the neighboring frames are same, then depth information can be passed to the neighboring frames with the help of motion vectors. This approach can reduce the computational burden to a significant extent in comparison with the conventional frame-by-frame approach. This approach paves the way for real-time implementation of fog removal algorithm. This chapter is structured as follows. In Section 6.2, challenges of real-time implementation are discussed. In Section 6.3, the video fog removal framework is explained in detail. Simulation and results are shown in Section 6.4. Section 6.5 concludes this chapter.

6.2 CHALLENGES OF REALTIME IMPLEMENTATION

A fast fog removal algorithm for image can be tried for the video fog removal by applying it on each frame of the foggy video. For real-time implementation, one needs to find the fastest algorithm with quality output. The execution time of the algorithm is investigated in Section 6.4.

For real-time implementation, the conventional frame by frame approach is computationally expensive. If temporal correlations between neighboring frames can be utilized for the restoration, it promises a large amount of saving in computation. Temporal redundancy removal is an essential step of any video coding standard (viz. MPEG)[A. M. Tekalp, 1995; S. Hoelzer, 2005]. In MPEG paradigm, Intra-frame (I frame) is spatially encoded and Predicted frames (P frame) & Bi-predicted frames (B frame) are temporally encoded followed by spatial encoding. Temporal encoding is performed using motion vectors. Estimation of depth information is the most computation-intensive step in the restoration of foggy image. As this step is iterative in nature, it is more difficult to implement in real-time. For the design of this step, an upper bound for the number of iterations is considered. According to the video coding standard, a GOP (Group of Pictures) has one I frame, and the rest are P and B frames. Computation can be reduced by estimating the depth information for the I frame only in the GOP. For P and B frames, depth map is estimated by the depth map previously calculated along with the motion vectors. With the use of temporal correlations, computation per frame can be reduced compared to conventional frame by frame approach. If image fog removal algorithm is fast then further improvement in speed can be achieved in video fog removal algorithm.

6.3   VIDEO FOG REMOVAL FRAMEWORK

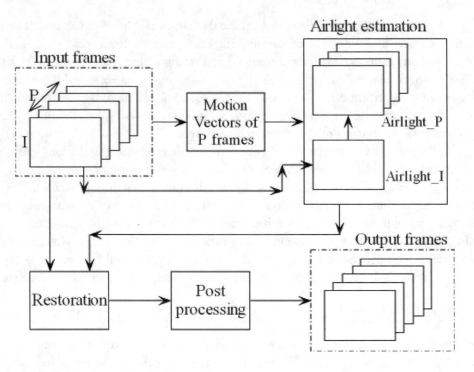

Figure 6.1: Block diagram of fog removal algorithm for video.

Block diagram of the presented fog removal algorithm is shown in Fig. 6.1. For ease of discussion, the GOP is assumed to consist of I and P frames only. Airlight map of I frame can be computed with any reported fog removal algorithm using single uncalibrated camera. Any algorithms mentioned in the literature survey and Chapter 5 can serve the purpose. In the fog removal scheme, post-processing step is the same as discussed in Section 5.2.5.

Motion vectors for P frames are estimated using the previous frames as reference. Here it is assumed that the airlight map remains the same for objects common in the frames within a GOP [A. M. Tekalp, 1995]. Hence, using the motion vector information the airlight map of the P frame under consideration can be reconstructed from the airlight map of the reference frame. After the estimation of airlight, the corresponding frame can be restored using airlight map information. In any fog removal algorithm for images captured by single uncalibrated camera system, airlight map estimation is the most computation-intensive step. In the presented video fog removal technique, computation is reduced with the use of and reference frame airlight map motion vectors. For GOP consisting of N frames (i.e., 1 I frame and N-1 P frames) using a conventional approach, airlight map need to be estimated for each frame. For this approach, computation is reduced to the estimation of airlight map once per GOP. However, the computation will slightly increase due to the computation of motion vectors and reconstruction of airlight map for each P frame. This technique can be easily extended for B frames.

6.3.1 MPEG CODING

MPEG bit stream consists of a series of coded frames. Coding a frame always starts by representing the original frame in YC_bC_r color space. The Y component represents luminance, and C_b and C_r components represent the chrominance. The color space transformation is useful in reducing the redundant information. There are three possible types of frames, called I, P and B frames. MPEG uses block based coding, which means that the frame is not encoded as a whole; it is divided into many independently coded blocks called macroblock. An I frame is an intra or spatially coded. It is an image compressed in a manner similar to a JPEG image. A P frame is inter or temporally coded, which means it uses the correlation between the current and a past frame for coding. Temporal coding is achieved using motion vectors [A. M. Tekalp, 1995]. There are few common algorithms for finding motion vectors such as:

1. **Sequential search** - It is a brute force method where every possible match is tested within a given search window (see Fig. 6.2(a)) [A. M. Tekalp, 1995].

2. **Logarithmic search** - Here, nine locations centered around the original macroblock are searched, then the search is repeated centered on the best match of the last iteration. At each iteration, the search window gets smaller until the desired fidelity is reached (see Fig. 6.2(b))[A. M. Tekalp, 1995].

3. **Hierarchical search** - The macroblock and search window are decimated once or more, and searching takes place in order of the lowest to the highest resolution so the motion vectors can be refined at every step [A. M. Tekalp, 1995; Yao et al., 2001] (see Fig. 6.2(c)).

   In B frame motion vectors are estimated not only from a previous frame, but also from a future frame, or both past and next frame.

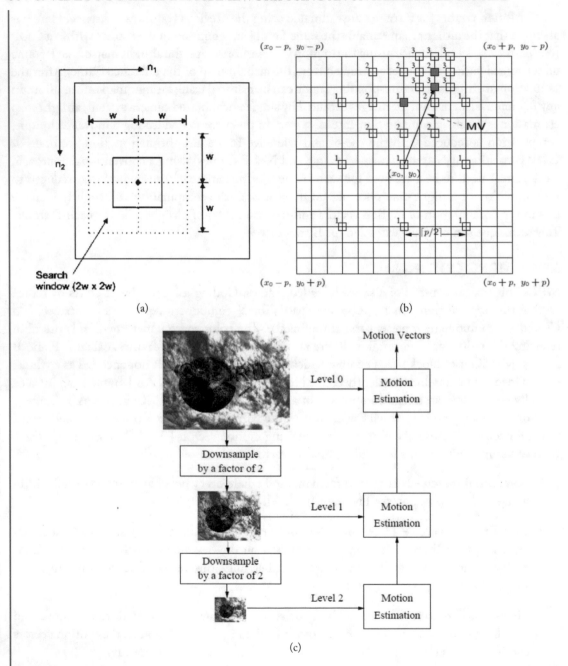

Figure 6.2: Common algorithms for motion vectors estimation (a) sequential search, (b) logarithmic search and (c) hierarchical search.

6.4   SIMULATION AND RESULTS

Table 6.1: Computation time (in sec) of competing fog removal algorithms on 400×600 image (Simulation platform MATLAB 7.0.4)

Method	t_{comp} (mean $\pm$ std.)
Fattal	$19.27 \pm 4.6 \times 10^{-3}$
Zhang et al.	$14.97 \pm 4.5 \times 10^{-3}$
He et al.	$14.35 \pm 4.2 \times 10^{-3}$
Tarel et al.	$4.12 \pm 5.1 \times 10^{-3}$
Presented (RGB)	$3.26 \pm 8.1 \times 10^{-3}$
Presented (HSI)	$3.12 \pm 7.9 \times 10^{-3}$

Table 6.2: Contrast gain (C_{gain}) and percentage of saturated pixels (σ) on image 'lonavala', 'ny17' and 'pumpkins'

Image	Method					
	Tarel et al		Presented (RGB)		Presented (HSI)	
	C_{gain}	σ	C_{gain}	σ	C_{gain}	σ
'lonavala'	1.19	0.00	4.32	0.00	4.79	0.00
'ny17'	6.62	0.0011	8.37	0.0001	12.71	0.00
'pumpkins'	3.33	0.00	5.58	0.00	6.91	0.00

Simulation is carried out in MATLAB 7.0.4 environment on a system with a 2.40 GHz Intel(R) Core(TM)2Duo CPU and 2 GB of RAM. For the simulation, the first task is to identify an efficient fog removal algorithm for images and then the same can be applied for fog removal of the video. The results for the computation time (t_{comp}) are shown in Table 6.1. Here each algorithm is executed ten times to generate the statistics. According to the execution time (t_{comp}), these algorithms can be arranged in ascending order of speed as: R. Fattal [2008], Zhang et al. [2010], He et al. [2009], J. P. Tarel and N. Hautiere [2009], presented (RGB) and presented (HSI). For video fog removal, three computationally efficient fog removal algorithms are chosen (Presented anisotropic diffusion-based algorithm using RGB and HSI color space and Tarel et al.). These algorithms can be good candidates for real-time implementation. Results show that the presented algorithms require less time in comparison with Tarel et al. Reason is that present (RGB) and (HSI) algorithms use anisotropic diffusion which require less time in comparison with linear operators used by Tarel et al. In between restoration in RGB and HSI color space, working in HSI color space requires less time. In HSI color space, algorithm works only on intensity and saturation components, while hue component remains unaltered, which explain result. Another

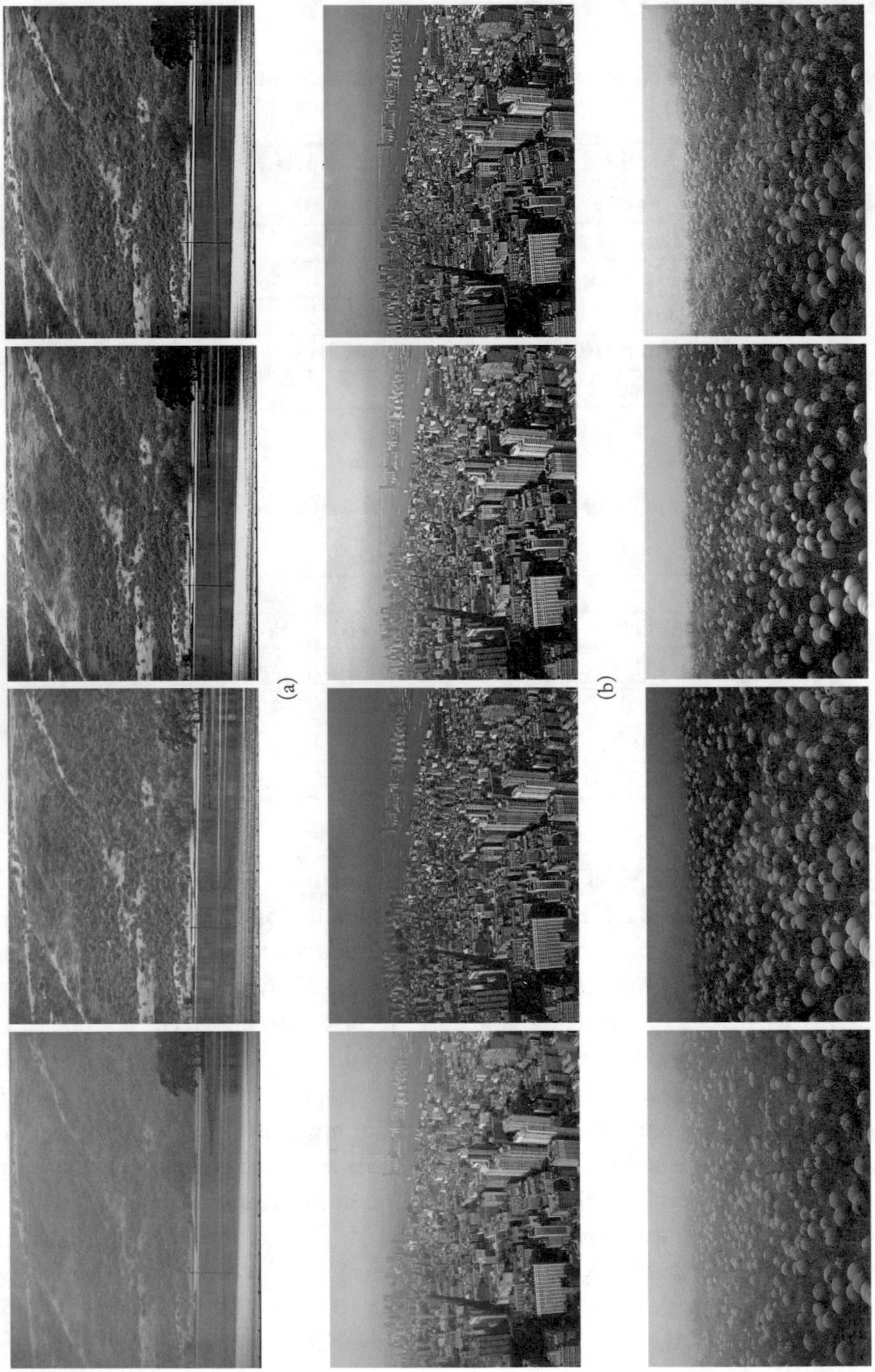

Figure 6.3:   The first column shows the original foggy (a) 'lonavala', (e) 'ny17' and (i) 'pumpkins' image. The second, third and fourth columns show the corresponding images restored by Tarel et al., Tripathi et al. (RGB), and Tripathi et al. (HSI), respectively.

benefit of working in HSI color space is that it maintains the color fidelity. The results verify that this algorithm is best for the real-time application. The results show that the quality of the restored images produced by the anisotropic diffusion-based algorithms in RGB and HSI color space are better than those produced by Tarel et al.

For quantitative analysis of the restored images, contrast gain (C_{gain}) and percentage of the number of the saturated pixels (σ) are chosen. Results are shown in Table 6.2. High contrast gain and low percentage of saturated pixels confirm the efficacy of the algorithm in HSI and RGB color space. From Table 6.2, we get that the anisotropic diffusion-based algorithm computed in HSI color space is the best among the three and closely followed by the anisotropic diffusion-based algorithm in RGB color space. The presented algorithm in RGB and HSI color space is closely followed by Tarel et al.

For video encoding, MPEG is an industry standard for the generic coding of moving pictures and associated information. The composition of GOP in MPEG standard is discussed in Section 6.2. In this simulation study, MPEG-2 simple profile is followed. In MPEG-2 simple profile, video is encoded in single I frame and couple of P frames per GOP. The P frames use the previous I frame or encoded P frame for temporal encoding while I frames are intra coded, i.e., the redundancy removal is limited to the spatial domain. For temporal redundancy removal, the motion vectors are estimated for 16×16 blocks. Motion vector estimation is performed by 'block matching' search in the reference frame. In the present work, logarithmic search with an initial step size of 8 is employed due to its merits. However, there can be other choices for the motion vector estimation. Here MPEG specified subsampling format 4:2:0 is used, which means the chrominance sampling is decimated by 2 in the horizontal and vertical direction. Both chrominance channels are reduced to one-quarter to the original; the net effect is total frame data rate to be cut in half without any perceptual effect on image quality. A macroblock is of 16×16 pixels and each macroblock is further divided into 8×8 pixels blocks. This results in 6 blocks per macroblock (4 for luminance and 2 for chrominance) [A. M. Tekalp, 1995; S. Hoelzer, 2005]. In MPEG standard, each block of I frame is processed independently with an 8×8 DCT. Temporal encoding in P frames is achieved using motion vectors. The basic idea is to match each macroblock in the past reference frame as closely as possible. Here, we are encoding only motion vectors not the real motion [K. E. Rapantzikos, 2002]. In the simulation, we have used block matching technique for the motion estimation. In block matching technique, assumption is that the pixels of a small image block exhibit the same motion from frame to frame. Therefore, the same motion vector is assigned to all pixels within the block. All motion analysis algorithms face few common problems such as motion discontinuities and temporal discontinuities. These problems are due to 3D to 2D projection of the real motion field. The airlight map depends on the depth. For the removal of the fog, this depth information is required. If there are no moving objects in the video, then depth information computed for the previous frame will remain the same for the next frame. If there are moving objects in the video then in successive frames its location and depth of the objects will change. This change in the location is estimated using the motion

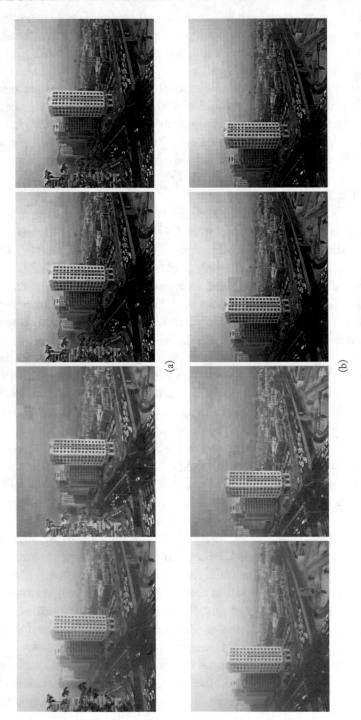

(a)

(b)

Figure 6.4: (a) 10th frame of 'abudhabi' video, restored by (b) Tarel et al., presented anisotropic diffusion-based restoration algorithm applied in (c) RGB color space and (d) HSI color space, (e) 60th frame of 'abudhabi' video, restored by (f) Tarel et al, presented anisotropic diffusion-based restoration algorithm working in (g) RGB color space and (h) HSI color space. (These are the 4th P frame of a (I+4P) GOP.)

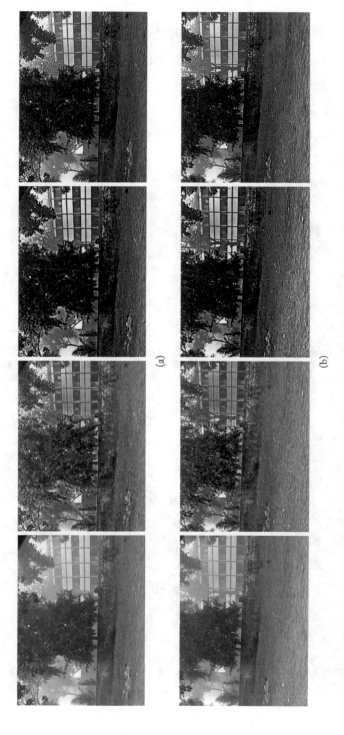

Figure 6.5: (a) 25th frame of 'fogiitcampus01' video, restored by (b) Tarel et al., presented anisotropic diffusion–based restoration algorithm applied in (c) RGB color space and (d) HSI color space, (e) 40th frame of 'fogiitcampus01' video, restored by (f) Tarel et al., presented anisotropic diffusion–based restoration algorithm working in (g) RGB color space and (h) HSI color space. (These are the 4th P frame of a (I+4P) GOP.)

Figure 6.6: (a) 90th frame of 'fogiitcampus02' video, restored by (b) Tarel et al., presented anisotropic diffusion–based restoration algorithm applied in (c) RGB color space and (d) HSI color space. (This is the 4th P frame of a (1+4P) GOP.)

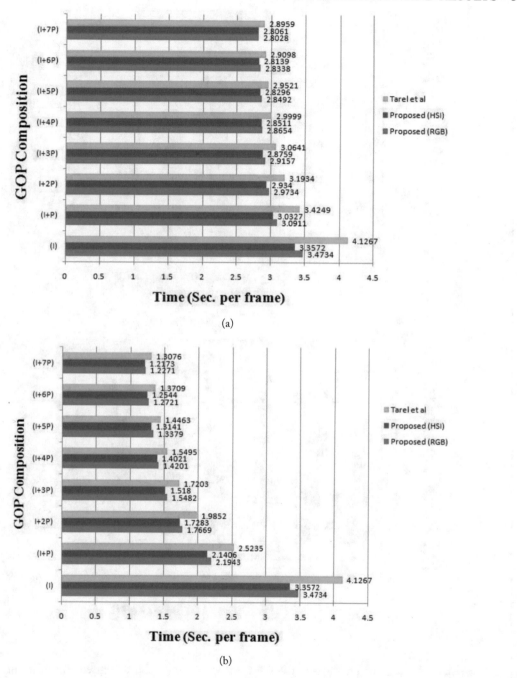

Figure 6.7: Computation time per frame for 'fogiitcampus01' video using different fog removal algorithms (a) with motion vectors computation (b) without motion vectors computation.

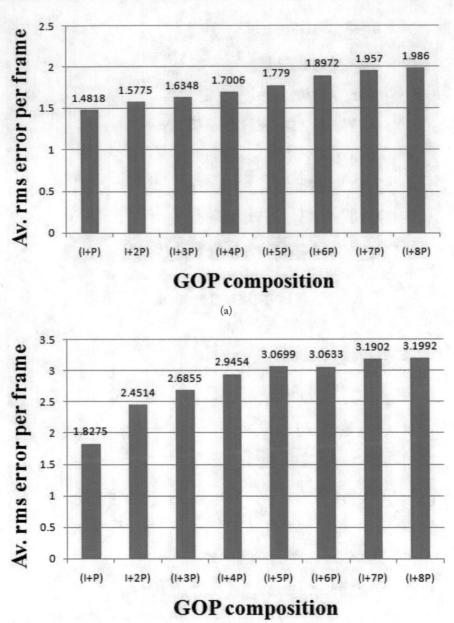

Figure 6.8: Average rms error per frame for restoration with the presented algorithm in RGB color space for varying the number of P frames in between I frames for (a) 'abudhabi' and (b) 'fogiitcampus01' video clips.

vectors of the object. Hence, in the present framework airlight map for the P frame is obtained with the help of motion vectors. Here it is assumed that the depth in a location is changed only due to the motion of the objects in the frame and the depth of the objects remains constant within the GOP.

Results of the video frames are shown in Figs. 6.4, 6.5 and 6.6. Results show that presented algorithms for video can effectively remove the fog. In Fig. 6.6, it can be seen that walking men are better visible after the removal of the fog, justifying the application of the presented framework along with the driver assistance system (DAS). It is shown that as the number of P frames increases in GOP, computation time per frame decreases (see Fig. 6.7) irrespective of the fog removal algorithms. In the present approach, fog removal algorithms for image are applied on I frame. For the fog removal of P frames, the required airlight map is estimated by the airlight estimate of the reference frame using motion vectors. For each frame (I & P), the procedure of restoration and post-processing is same after the airlight map estimation. Hence, for P frames computation is required only for the motion vector estimation and restoration (see Fig. 6.7(a)). Irrespective of the fog removal algorithms, computation per frame for a GOP of size 3, 5 and 8 is reduced by 13–23%, 15–28% and 16–30% respectively. However, these motion vectors can be obtained by the encoder/decoder if the system is used as video pre/post-processing step. If these motion vectors are available then computation is further reduced (see Fig. 6.7(b)). In this case, irrespective of the fog removal algorithms, computation per frame for a GOP of size 3, 5 and 8 is reduced by 48–52%, 58–62% and 64–68% respectively. The availability of motion vectors reduces the computation significantly which facilitates real-time application of fog removal algorithms. The results shown in Fig. 6.7 are averaged over 50 frames. In the fog removed videos, visual quality of each frame also depends upon the GOP composition (i.e., number of P frames between I frames). Appropriate choice of the number of P frames in GOP can be made to maintain low RMS error in reconstruction. The RMS error is computed with respect to the frame restored as I frame and low value of RMS shows the better performance. It is noted that as the number of P frames between two I frames increases, visual quality of frames decreases (see Fig. 6.8). Results are shown for the average of 30 frames for the presented (RGB) algorithm. Similar results can be observed for the presented (HSI) algorithm.

Perceptual quality of the restored image is judged in terms of the perceptual quality metric (PQM). Qualitative results are shown in Fig. 6.9 and their corresponding PQM scores are shown in Table 6.3. Results show that PQM score is close to 10, and perceptually there is no difference between the frame by frame approach and presented video framework approach. Results in Fig. 6.8(b) (RMS error) and Table 6.3 (PQM score) have a negative correlation of value -0.0876, which justify that as the RMS error increases PQM score decreases. Thus, it can be said that appropriate choice of the number of P frames paves the way for the real-time implementation of the fog removal algorithm without perceptual loss of visual quality of restored video. More qualitative results can be found at `http://www.ecdept.iitkgp.ernet.in/web/faculty/smukho/docs/fog_video/fog_video.html`.

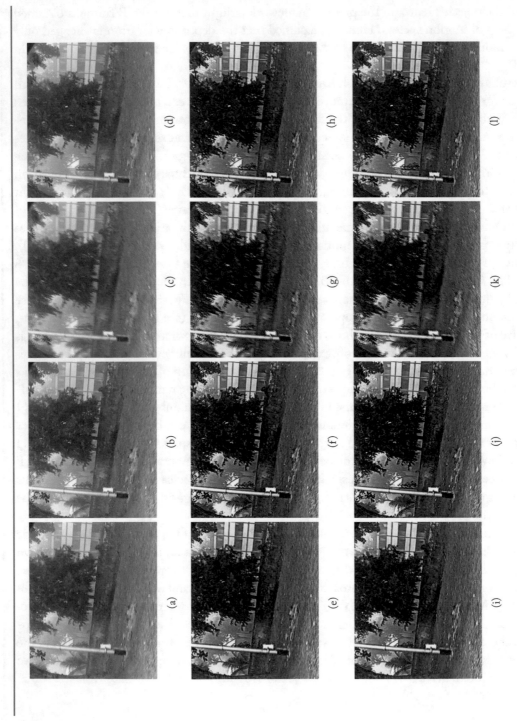

Figure 6.9: For 'fogiitcampus01' video: (a)-(d) different original foggy frames, (e)-(h) frames restored using frame by frame approach, frame restored by presented video framework treating as (i) 3rd P frame, (j) 5th P frame, (k) 8th P frame, (l) 9th P frame (GOP size 9 is considered here and the results are shown for the presented RGB algorithm).

Table 6.3: Perceptual quality metric (PQM) score of P frame of video 'fogiitcampus01' of restoration using the presented algorithm in RGB color space. The GOP size of nine frames are used and the reported result is the statistics of three GOPs

n^{th} P frame	PQM Score (mean $\pm$ std.)
1	8.8944 ± 0.2237
2	8.8558 ± 0.1127
3	8.9530 ± 0.1977
4	9.0405 ± 0.2288
5	8.8283 ± 0.0957
6	8.9140 ± 0.1424
7	8.9032 ± 0.2170
8	8.7741 ± 0.1799

6.5   CONCLUSION

In this chapter, novel, efficient and fast fog removal algorithm for video is presented. Here it is claimed that any fog removal algorithm can be extended for video by making use of temporal correlation present among frames. Airlight map for I frame can be computed using any image fog removal algorithm. For P/B frames the airlight estimate can be derived from the reference frames using the motion vectors information. Results confirm that there is a significant improvement in the speed of video fog restoration. Theoretically, this technique can be applied until there is a scene change. On the other hand, the increase in the number of P/B frames increases the degradation of restored image. In simulation, it is noticed that average rms error increases with the increase in number of P/B frames in GOP but the PQM score is not significantly changed as long as the size of the GOP is within reasonable limit. Low degradation in reconstruction can be obtained with the appropriate choice of the number of P/B frame in GOP. Hence, it may be concluded that with the use of temporal correlation, computation per frame significantly reduced irrespective of the algorithms, which paves the way for real-time implementation of fog removal along with video encoding or decoding.

CHAPTER 7

Conclusions and Future Directions

This book leads us to efficient algorithms for fog from images and videos. The book started with the motivation for the development of fog removal algorithm. In Chapter 2, the atmospheric condition referred as fog is analyzed. Chapter 3 has described the sources where the foggy images and videos are available and the metrics used to measure the efficacy of the fog removal techniques. In Chapter 4, the development of fog removal algorithms for images and videos are described. In Chapter 5, a novel and efficient fog removal algorithm is presented for images captured by a single, uncalibrated camera system. Fog formation is due to attenuation and airlight. Attenuation reduces the contrast and airlight increases the whiteness in the scene. The presented algorithm uses anisotropic diffusion to recover scene contrast. Simulation results demonstrate that a presented anisotropic diffusion-based algorithm outperforms prior state of the art algorithms in terms of contrast gain, percentage of the number of saturated pixels and computation time. The presented algorithm is independent of the density of the fog and does not require user intervention. It can handle color as well as gray images. Along with the RGB color space, the presented algorithm can work on HSI color space which further reduces the computation.

In Chapter 6, a framework for real-time video processing for fog removal using uncalibrated single camera system is presented. Intelligent use of temporal redundancy present in video frames paves the way for real-time implementation. Any fog removal algorithm for images acquired with uncalibrated single camera system can be extended to video using the presented framework. For the purpose of real-time implementation, several fog removal algorithms for images are investigated, and a few top-ranking ones in speed and quality are chosen. Simulation results confirm that the presented framework reduces the computation per frame significantly.

Removal of fog from a single image is always an under constraint problem due to the absence of depth information. Hence, single image fog removal requires an assumption or prior. During the restoration of foggy image, it is necessary that both the luminance and chrominance should be recovered well to maintain the color fidelity and appearance. Hence, future research will focus on better estimation of depth information and restoration with better visual quality. A fast and accurate estimation of depth information increases the speed and perceptual image quality.

Consideration of wavelengths other than the visible range may be beneficial for the visibility of the scene in the bad weather (foggy) condition. Hybrid image processing algorithm, which will use multiple wavelengths—i.e., SWIR, MWIR, LWIR and radar—may open up a new direction of research in this field.

Bibliography

R. C. Gonzalez and R. E. Woods, *Digital Image Processing,* Addison-Wesley, Reading, Mass., 1992. 13, 24

K. Garg and S. K. Nayar, "Vision and Rain", International Journal of Computer Vision, Vol. 75, No. 1, pp. 3–27, 2007. DOI: 10.1007/s11263-006-0028-6. 2, 5

S. G. Narasimhan and S. K. Nayar, "Vision and the Atmosphere", International Journal of Computer Vision, Vol. 48, No. 3, pp. 233–254, 2002. DOI: 10.1023/A:1016328200723. 3, 6, 16

A. K. Tripathi, S. Mukhopadhyay and A. K. Dhara, "Performance Metrics for Image Contrast", International Conference on Image Information Processing, Shimla, India, Nov. 2011. DOI: 10.1109/ICIIP.2011.6108900. 3

A. M. Tekalp, *Digital Video Processing,* First Edition, Prentice Hall of India, 1995. 47, 49, 53

W. Yao, J. Ostermann, and Ya-Qin Zhang, *Video Processing and Communications,* First Edition, Prentice Hall of India, 2001. 49

S. Hoelzer, *MPEG-2 Overview and MATLAB Codec Project,* University of Illinois at Chicago (UIC), April 2005. 47, 53

K. E. Rapantzikos, "Dense Estimation of Optical Flow in the Compressed Domain Using Robust Techniques", M. Sc. Report, Department of Electronic & Computer Engineering, Technical University of Crete, September 2002. 53

Y. Y. Schechner, S.G. Narasimhan, and S.K. Nayar, "Instant Dehazing of Images Using Polarization", IEEE Computer Society Conference on Computer Vision and Pattern Recognition, pp. 325–332, 2001. DOI: 10.1109/CVPR.2001.990493. 6, 14, 15

S. G. Narasimhan and S. K. Nayar, "Chromatic Framework for Vision in Bad Weather", IEEE Conference on Computer Vision and Pattern Recognition, Vol. 1, pp. 598–605, 2000. DOI: 10.1109/CVPR.2000.855874. 13, 14

J. P. Tarel and N. Hautiere, "Fast Visibility Restoration from a Single Color or Gray Level Image", IEEE International Conference on Computer Vision, pp. 2201–2208, 2009. DOI: 10.1109/ICCV.2009.5459251. 8, 14, 19, 23, 40, 47, 51

R. Fattal, "Single Image Dehazing", International Conference on Computer Graphics and Interactive Techniques Archive ACM SIGGRAPH, pp. 1–9, 2008. DOI: 10.1145/1360612.1360671. 8, 13, 19, 40, 51

R. T. Tan, "Visibility in Bad Weather from a Single Image", IEEE Conference on Computer Vision and Pattern Recognition, pp. 1–8, 2008. DOI: 10.1109/CVPR.2008.4587643. 18

J. Kopf, B. Neubert, B. Chen, M. Cohen, D. Cohen-Or, O. Deussen, M. Uyttendaele and D. Lischinski, "Deep Photo : Model-Based Photograph Enhancement and Viewing", ACM Transactions on Graphics, Vol. 27, No. 5, pp. 116:1–116:10, 2008. DOI: 10.1145/1409060.1409069. 17

K. He, J. Sun and X. Tang, "Single Image Haze Removal Using Dark Channel Prior", IEEE International Conference on Computer Vision and Pattern Recognition, pp. 1956–1963, 2009. DOI: 10.1109/TPAMI.2010.168. 8, 9, 19, 25, 26, 40, 51

D. Kim, C. Jeon, B. Kang and H. Ko, "Enhancement of Image Degraded by Fog Using Cost Function Based on Human Visual Model", IEEE International Conference on Multisensor Fusion and Integration for Intelligent Systems, pp. 163–171, 2008. DOI: 10.1109/MFI.2008.4648109. 18

P. Perona and J. Malik, "Scale Space and Edge Detection Using Anisotropic Diffusion", IEEE Transactions on Pattern Analysis and Machine Intelligence, Vol. 12, No. 7, pp. 629–639, 1990. DOI: 10.1109/34.56205. 26, 27

C. Tomasi and R. Manduchi, "Bilateral Filtering for Gray and Color Images", Proceeding of the 1998 IEEE International Conference on Computer Vision, Bombay, India, 1998. DOI: 10.1109/ICCV.1998.710815.

T. L. Economopoulosa, P. A. Asvestasa and G. K. Matsopoulos, "Contrast Enhancement of Images Using Partitioned Iterated Function Systems", Image and Vision Computing, Vol. 28, No. 1, pp. 45–54, 2010. DOI: 10.1016/j.imavis.2009.04.011. 9, 10

N. Hautiere, J. P. Tarel, D. Aubert and E. Dumont, "Blind Contrast Enhancement Assessment by Gradient Ratioing at Visible Edges", Image Analysis & Stereology Journal, Vol. 27, No. 2, pp. 87–95, 2008. DOI: 10.5566/ias.v27.p87-95. 9

D. J. Jobson, Z. Rahman and G. A. Woodell, "A Multi-Scale Retinex for Bridging the Gap Between Color Images and the Human Observation of Scenes", IEEE Transactions on Image Processing: Special Issue on Color Processing, Vol. 6, No. 7, pp. 965–976, July 1997. DOI: 10.1109/83.597272. 13

S. G. Narasimhan and S. K. Nayar, "Shedding Light on the Weather", International Conference on Computer Vision and Pattern Recognition, pp. 665–672, 2003. DOI: 10.1109/CVPR.2003.1211417.

S. G. Narasimhan and S. K. Nayar, "Contrast Restoration of Weather Degraded Images", IEEE Transaction on Pattern Analysis and Machine Intelligence, Vol. 25, No. 6, pp. 713–724, June 2003. DOI: 10.1109/TPAMI.2003.1201821. 6, 16

S. G. Narasimhan and S. K. Nayar , "Removing Weather Effects from Monochrome Images", International Conference on Computer Vision and Pattern Recognition, pp. 186–193, 2001. DOI: 10.1109/CVPR.2001.990956. 16, 17

S. G. Narasimhan and S. K. Nayar, "Interactive (De) Weathering of an Image Using Physical Models," IEEE Workshop on Color and Photometric Methods in Computer Vision, In Conjunction with ICCV, October, 2003. 17

J. P. Oakley and B. L. Satherley, "Improving Image Quality in Poor Visibility Conditions Using a Physical Model for Contrast Degradation", IEEE Transaction on Image Processing, Vol. 7, No. 2, pp. 167–179, 1998. DOI: 10.1109/83.660994. 13, 15

J. P. Oakley and H. Bu, "Correction of Simple Contrast Loss in Color Images", IEEE Transaction on Image Processing, Vol. 16, No. 2, pp. 511–522, 2007. DOI: 10.1109/TIP.2006.887736. 18

J. Zhang, L. Li, G. Yang, Y. Zhang and J. Sun, "Local Albedo-Insensitive Single Image Dehazing", The Visual Computer, pp. 761–768, 2010. DOI: 10.1007/s00371-010-0444-z. 20, 40, 51

S. Fang, J. Zhan, Y. Cao and R. Rao, "Improved Single Image Dehazing Using Segmentation", IEEE International Conference on Image Processing (ICIP), 2010, pp. 3589–3592. DOI: 10.1109/ICIP.2010.5651964. 20

A. Levin, D. Lischinski and Y. Weiss, "A Closed Form Solution to Natural Image Matting", CVPR, vol. 1, pp. 61–69, 2006. DOI: 10.1109/CVPR.2006.18. 19

T. C. S. Azevedo, J. M. R. S. Tavares, M. A. P. Vaz, "3D Object Reconstruction from Uncalibrated Images Using an Off-the-Shelf Camera", Advances in Computational Vision and Medical Image Processing: Methods and Applications, ISBN: 978-1-4020-9085-1, DOI: $10.1007/978 - 1 - 4020 - 9086 - 87$, pp. 117–136, Springer, 2008. DOI: 10.1007/978-1-4020-9086-8_7. 17

National Highway Traffic Safety Administration (http://www.nhtsa.gov/) 2

Z. Wang, A. C. Bovik, H. R. Sheikh and E. P. Simoncelli, "Image Quality Assessment: From Error Visibility to Structural Similarity", IEEE Transactions on Image Processing, Vol. 13, No. 4, pp. 600–612, Apr. 2004. DOI: 10.1109/TIP.2003.819861.

Z. Wang, H. R. Sheikh, and A. C. Bovik, "No-Reference Perceptual Quality Assessment of JPEG Compressed Images", Proc. IEEE International Conference on Image Processing, vol. 1, pp. 477–480, 2002. DOI: 10.1109/ICIP.2002.1038064. 9, 11

C. Rouvas-Nicolis and G. Nicolis, "Stochastic Resonance", Scholarpedia, Vol. 2, No. 11, pp. 1474, 2007. 11

P. Perez, "Markov Random Fields and Images", CWI Quarterly, Vol. 11, No. 4, pp. 413–437, 1998. 19

R. Kindermann and J. L. Snell, *Markov Random Fields and Their Applications*, American Mathematical Society, First Edition, 1980. DOI: 10.1090/conm/001. 19

Authors' Biographies

SUDIPTA MUKHOPADHYAY

Sudipta Mukhopadhyay is currently Associate Professor in the Electrical and Electrical Communication Engineering, IIT Kharagpur. He received his B.E. degree from Jadavpur University, Kolkata, in 1988. He has received his M.Tech. and Ph.D. degrees from IIT Kanpur in 1991 and 1996, respectively. He served several companies including TCS, Silicon Automation Systems, GE India Technology Centre and Philips Medical Systems before joining IIT Kharagpur in 2005 as Assistant Professor of Electrical and Electrical Communication Engineering, IIT Kharagpur. In 2013 he become Associate Professor in the same department. He has authored or co-authored more than 70 publications in the field of signal and image processing. He has filed seven patents while working in industry and continued the trend after joining academia. He is a senior member of the Institute of Electrical and Electronics Engineers (IEEE), Member of SPIE and corresponding member of Radiological Society of North America (RSNA). He has done many applied projects sponsored by DIT, Intel and GE Medical Systems IT, USA. He is also founder and director of Perceptivo Imaging Technologies Private Ltd., a company under the guidance of S.T.E.P. IIT Kharagpur. The company specializes in developing innovative software for signal and image processing.

ABHISHEK TRIPATHI

Abhishek Tripathi is currently working as Senior Engineer at Uurmi Systems Pvt. Ltd., Hyderabad, India. He received his Ph.D. degree from Indian Institute of Technology Kharagpur, India, in 2012. He received the M.Tech. degree from National Institute of Technology, Kurukshetra, India, in 2008. He received the B.Tech. degree from Uttar Pradesh Technical University, Lucknow, India, in 2006. His research interests include computer vision, image-based rendering, nonlinear image processing, physics-based vision, video post-processing, p recognition, machine learning and medical imaging.

Printed in the United States
by Baker & Taylor Publisher Services